How Scientists Communicate

How Scientists Communicate

DISPATCHES FROM THE FRONTIERS OF KNOWLEDGE

Alan Kelly

OXFORD
UNIVERSITY PRESS

OXFORD

UNIVERSITY PRESS

Oxford University Press is a department of the University of Oxford. It furthers the University's objective of excellence in research, scholarship, and education by publishing worldwide. Oxford is a registered trade mark of Oxford University Press in the UK and certain other countries.

Published in the United States of America by Oxford University Press
198 Madison Avenue, New York, NY 10016, United States of America.

Library of Congress Cataloging-in-Publication Data
Names: Kelly, Alan (Alan L.), author.
Title: How scientists communicate : dispatches from the frontiers of knowledge / Alan Kelly.
Description: New York, NY : Oxford University Press, [2020] |
Includes bibliographical references and index.
Identifiers: LCCN 2019053119 (print) | LCCN 2019053120 (ebook) |
ISBN 9780190936600 (hardback) | ISBN 9780197521021 (epub)
Subjects: LCSH: Communication in science. | Technical writing.
Classification: LCC Q223 .K45 2020 (print) | LCC Q223 (ebook) |
DDC 808.06/65—dc23
LC record available at https://lccn.loc.gov/2019053119
LC ebook record available at https://lccn.loc.gov/2019053120

1 3 5 7 9 8 6 4 2

Printed by Sheridan Books, Inc., United States of America

For Louis and Ena Kelly

{ CONTENTS }

{ ACKNOWLEDGMENTS }

One of the themes of this book is about how science and scientific publication are typically today a collaborative effort, and in this spirit I wish to acknowledge several individuals who helped during the writing of this book.

First, I thank Jeremy Lewis, my editor at Oxford University Press, for his positive reaction to my proposal and subsequent support, advice, and patience throughout the writing and editing process. I also thank Suganya Elango of Newgen Knowledge Works and Bronwyn Geyer at Oxford University Press, for help during the preparation of the book for publication, and for the expert copyeditors whose input into the final manuscript was very helpful. Great thanks also to Brían French, for his creative collaboration on cover design concepts.

I thank friends and colleagues who provided perceptive feedback on draft chapters, including Dr. John Finn at Teagasc, Professor Thom Huppertz from Wageningen University Research and FrieslandCampina, and Dr. Fergus McAuliffe from University College Dublin, as well as many scientific colleagues and collaborators whose inputs and comments have informed my views presented here (especially Hugh Kearns from Flinders University, Australia), and my brother, Paul Kelly.

I also thank all the students and researchers who have attended my workshops and courses on scientific writing in Ireland for over a decade (especially the long-running module PG6001, or STEPS), whose feedback showed me how critical development of good communication skills is, and whose questions and in-class discussions led to many of the points discussed here. I further thank a number of these who provided feedback on draft chapters, especially Colm Breathneach of the Tyndall National Institute, and members of my own Milk Science and Technology research team and the Food Ingredients Research Group at University College Cork, who helped by providing feedback on drafts and checking proofs. I also acknowledge that most of this book was written on a short sabbatical from University College Cork in Ireland in 2018, and recognize the support of the College of Science, Engineering and Food Science and the School of Food and Nutritional Sciences for facilitating this.

Finally, thanks as always to my wife, Brenda, and my children, Dylan, Martha, and Thomas, for their support and encouragement during the writing of this book,[1] and for tolerating my being around the house more than usual while doing so,[2] as well as my extended family. I dedicate this book to the memory of my late parents.

---

[1] Special mention for Martha's indexing skills!

[2] I also recognize the nonhuman support of Juno and Odie, who taught me that it is actually just about possible to write with a dog curled up on your lap.

{ ABOUT THE AUTHOR }

Alan Kelly is a Professor in Food Science and Technology at University College Cork in Ireland, where he specializes in teaching and research in the science of dairy products, and food processing and innovation more generally. He also is very interested in, and regularly delivers courses and workshops in, all forms of scientific communication, and is the author of *Molecules, Microbes and Meals: The Surprising Science of Food*. He lives in Cork with his wife, Brenda; three children; and two dogs.

Introduction

Let's start with a simple question: What do scientists actually do?

In most cases, they do research, the goal of which is to learn more about the world in all its aspects, whether the topic is our own bodies, the smallest particles that make up matter, or the vastest reaches of the universe.[1] Their research goal may be to fight disease, feed the world, create new technologies, understand our climate, or any of a million other objectives specific to different areas and disciplines.

The point of all this research then is to add to our storehouse of human knowledge, whether with practical consequences in mind or sometimes for the goal of simply "understanding more."

We see the outputs and benefits of this research all around us every day, in medicine, technology, food, communications, and countless other facets of our science-filled lives, and we can read about our state of knowledge in books, websites, and articles. However, behind every achievement, benefit, fact, theory, or argument, seldom seen or appreciated, there are the scientists whose work has given rise to it.

Science is a fundamentally human endeavor, driven by the hard work, curiosity, commitment, and ambition of researchers, and sometimes complicated by human factors like jealousy, competitiveness, insecurity, and (rarely, we hope) dishonesty. Through the work of scientists, experiments are undertaken, trials designed, fieldwork completed, and data measured, analyzed, and pored over to extract new learning about the topic in question. This learning then, if validated and found to be reliable and of benefit, becomes part of the book of knowledge.

We could picture the world of science and research as a sphere, near to which is found a much larger sphere, which represents the world of impact, whether through applications in everyday life or just the bank of knowledge that exists and can be drawn on, admired, or learned. In between these two spheres is a narrow channel, through which information flows from the world of research to the real world (as illustrated schematically in Fig. 1.1).

---

[1] Exceptions being where scientists work in industry in production, analysis, or other areas.

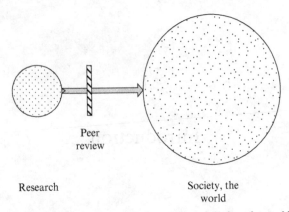

FIG. 1.1 *The relationship between the world of research and the broader world and modern society, linked by a flow of information that is policed by the process of peer review.*

In this channel there is a checkpoint or gate, which is a tough barrier to get past.[2] Information that seeks to cross over is scrutinized, questioned, and analyzed, and much gets turned back. This then either never enters the public sphere or has to find less reputable back channels, arriving on the fringes of the real world as shadows or unreliable messages.

That which passes the gatekeepers, though, is stamped with authority and credibility and becomes part of the world, becomes part of what we know, and adds to the databank of our understanding of the world.

Those whose names are associated with providing that information are not ignored either, and their association with such successful messages from the world of science into society is recognized and brings them benefits in terms of their standing and future careers.

In this model (simplistic of course, as the sphere of research sits within, not beside, the sphere of the world, but forgive me the analogy to avoid topological complications!), the bridge is scientific communication, and the gate symbolizes the standards and processes that science has evolved over the last couple of centuries to manage and assure the quality of the advice that the scientific community provides to society.

For science to relate to and impact on the world depends completely on the effectiveness by which scientists can communicate their work and its importance. For this reason, communication skills are arguably the most important skills researchers should have in their professional toolkit.

If we consider that the objective of science is to add to the store of human knowledge, the passing on of knowledge may have long ago been an oral tradition

---

[2] For some reason, I picture here the keeper of the Bridge of Death in *Monty Python and the Holy Grail* asking travelers the average air-speed velocity of an unladen swallow.

but, since science has become a sophisticated professional activity, it has been a written history. The writer Anne Sayre described it thus:

> Scientists are communicative people, they are obliged to be. The obligation also exists in the sense of a duty requiring them to publish their findings. In science, even more than elsewhere, to suppress the truth is to consent to a lie. If half the motive behind the duty to publish one's findings is a duty toward oneself, the other half is an acknowledgement of the necessity of pooling information and knowledge for the sake of science itself. Indeed, if this were not done, and new truths were kept secret, the progress of science would slow to a crawl, if only because time and energy and the resources of intellect would be devoted to repeating what has already been done, rediscovering what has been discovered, duplicating what already exists.[3]

Scientists need to be able to explain, argue, enthuse, and convince, at many different levels, and need to master the production of the gold standard format of scientific communication—a formalized account of exactly why and how a piece of research was done, what was found, and what it means: the *peer-reviewed scientific paper.*

The scientific paper is perhaps the most important part of modern civilization and society that is invisible or unappreciated by most people; it is as simple and as complex as that. It is easy to sound pompous or as if I am trying to reinforce the intellectual superiority of the scientific enterprise, but the scientific paper has been the way in which mankind has recorded its knowledge for several centuries. It is the means by which those who have dedicated their professional activities to science get to evaluate and assimilate the results of the knowledge generated by their colleagues, knowing that this has been subjected to rigorous screening to (hopefully) ensure the reliability of what is reported. The scientific paper has thereby evolved into a highly structured way of presenting the output of scientific research in a way that serves both writer and reader best.

Every individual scientific paper is an entry into the ever-changing larger entity that is called the *scientific literature*, which comprises tens of thousands of individual periodicals called *scientific journals*. These, not that long ago, comprised millions of pages of paper in libraries around the world, but today the literature is largely electronic, with vast amounts of knowledge being accessible instantly to anyone with the necessary tools at their fingertips.[4]

The scientific literature is essentially the encyclopedia of our knowledge about the world and ourselves. Unlike other encyclopedias, however, the scientific literature does not undergo periodic revisions, between which times it remains static, slowly becoming outdated in areas where knowledge is rapidly changing. Instead, it is a living and constantly changing storehouse of facts, a book that continuously

---

[3] From Sayre, Anne. *Rosalind Franklin and DNA.* (New York: W.W. Norton and Co., 1975), p110.

[4] Although, as we will see, this can be complicated by the presence of paywalls behind which much of what is published is actually found.

writes itself, thanks to the inputs of millions of researchers and scientists around the world.

Scientific publication is a process that begins when a scientist or group of scientists decides that they have something to say which their peers in the scientific community should know about, and that should have an interest or impact beyond the scientific community. They then document what they have to report, in a stylized and regulated format, and send it to their chosen medium of communication, or scientific journal. In a scientific equivalent to someone tapping a wine glass with a knife at a public event, the act of writing and submitting for publication a paper gives notice that the writer(s) have something to say and that others should listen.

The scientific community is intrinsically skeptical of what their own have to say, however, due to their being trained to question and challenge that which they see and hear. No matter the preeminence of the writers, the process of *peer review* (the ferocious gatekeeper on our metaphorical bridge) springs into action to determine whether the submitted paper is worthy of publication in the journal to which it was submitted.

Not all journals are equal, and there are clear hierarchies and leagues of prestige; acceptance in the top journals in the world carries great reward in terms of recognition and credit. The flip side of this is that the possibility of rejection and the expected quality and importance of the paper being considered by a top journal are proportionately higher.

So science, fundamentally, is communicative. Scientists may say they do science for themselves, to satisfy their own need to know, but if they do not publish it, very few will believe they have got the full satisfaction out of the work. It is not enough just to know, it is necessary for scientists to go to the rooftops or the mountain peaks and shout it out too. They are saying, in part, "Look what I found," and, certainly also in part, "Aren't I bloody clever to have found it?" The second motivation serves the self, the first serves the ultimate master: science itself.

Why Do Researchers Publish Papers?

Throughout the chapters of this book, we will explore several fundamental questions about the publication process, including the following:

- Why do researchers publish?
- Where do they publish?
- How do they prepare to publish?
- What determines whether their work gets published?
- What happens after publication?
- What other means are available to disseminate scientific knowledge?

This sequence of quite logical questions forms the architecture of this book, as the following chapters look at each of these questions.

We will consider here, though, the first of these questions: Why do researchers publish?

To answer this, let's back up a little and ask instead: what does a researcher do? The clear and not intentionally facetious answer is that, in any field, they do research, by which I mean that they seek to uncover new knowledge. In the case of scientific research, this means broadly the discovery or explanation of aspects of the world, from the smallest inner components of atoms to the largest structures in the universe, passing through every molecule and life form in between.

Having generated such knowledge, the first step has been completed, but the research will only have an impact and significance if it is communicated. The Irish writer John McGahern, when asked when he knew a book was finished, said that he didn't finish it, the reader does. A paper is like that, in that the process is completed when the outcome has been read and acted on by those who represent the target audience.

Typically, a scientist (or group of scientists) will do a piece of research that hopefully answers a useful question and provides interesting outcomes that someone would find useful, and presumably perform their work in a reliable and careful manner such that the results could be trusted as reliable. However, imagine if for some reason they decided not to tell anyone about this. What would be the benefit, other than perhaps a warm glow of pride, if they decided not to share their findings with anyone?

Now there might be good (or not so good) reasons why they decide not to share this, or at least not share it immediately. One may be that need to do more follow-up research before publication, or that the objective was not originally that the outcomes be widely shared, for example if commercial concerns such as applying for a patent are involved. In more problematic scenarios, which we will touch on in more detail in chapter 8, they may be prevented or discouraged from publishing due to the wishes of other parties, such as funding agencies or even governments, if the results uncovered do not fit with what these bodies wish to have in the public domain.

There may also be arguments around whether certain types of research are unsafe to publish, in terms of providing highly detailed information that, as well as being used for good, could be used for terrible ends. For example, in 2011 researchers in the Netherlands encountered significant (and well-founded) expressions of concern when they proposed to publish a paper on how the transmission of avian flu to humans (and hence the danger posed by the virus) could be increased through genetic mutation. Such research, on the one hand, was critical for medical research, but (it was argued) could be twisted to present a blueprint for a bioterrorism scenario too horrific to warrant publication without careful consideration of the possible dangers.[5]

---

[5] The research was eventually cleared for publication, but prompted intense and ongoing debate about whether the open nature of scientific publishing should in some cases be modified to allow critical

In a different variant on this moral dilemma, a *New York Times* article in 2018[6] described a researcher finding themselves in a very difficult position when preparing to publish the results of research they had conducted into the relationship between vaccination for influenza and the immunity of those vaccinated. The outcomes were believed by other researchers in the field to cause worry about the impact of vaccines at a time when "anti-vax" campaigners were creating doubt in the public mind about these medicines without scientific basis. The researchers pushed ahead, believing that the results were so important that they had no right not to publish them, and their work underwent at least eight reviews, a very rare degree of intense scrutiny, before being published.

However, assuming such concerns do not apply in our current hypothetical situation, if the researcher or researchers somehow chose not to publish their work, it could again reasonably be asked what the point of doing the research was. For this reason, we could claim that research is not complete until it has been recorded and passed on to those who should know the outcomes of that research.[7]

Science has been compared to a wall, with every piece of research being like a brick that adds to that wall, and later research fitting in around and on top of the brick, and this is a good analogy. Isaac Newton once said, "if I have seen further it is by standing on the shoulders of giants"—a lovely[8] metaphor for how researchers build on the successes of others. The foundations or shoulders for others to build or stand on are the papers of those that came before.

There are also deeply personal reasons for scientists to publish papers. Scientists unquestionably measure status, and others evaluate their status, not just by the length of their publication lists but also by the quality of the publications therein, as assessed by a range of parameters that will be discussed in later chapters of this book. In terms of how others evaluate scientists' status, the brutal truth is that, when it comes to promotion, jobs, establishment, or tenure and many other career-defining considerations for a modern scientist, publishing is really a matter of survival, reflected in the unfortunately accurate cliché of "publish or perish."

---

details to be withheld (as was practiced in this case), to prevent such research being essentially weaponized for nefarious purposes.

[6] "Anti-vaccine activists have taken vaccine science hostage" *New York Times*, August 9, 2018.

[7] This raises the question of whether a paper that subsequently appears behind a publisher's paywall can really be said to be available to all those who should know the outcome of the research, an issue we will come back to in later chapters.

[8] But not an uncontroversial one! Newton wrote this in a letter to fellow scientist and rival Robert Hooke, a man of apparently small stature, and some have proposed undertones of sarcasm in terms of the physical or intellectual heights of those from whom he obtained his inspiration. The British band Oasis named an album *Standing on the Shoulder of Giants*, which is an anatomically confusing twist on the line.

A Brief Biographical Note

It may be useful, before leaving this Introduction section and moving into the chapters to follow, for readers to appreciate the perspective from which I have written this book.

My field is food science, which relates to understanding the ways in which raw food materials are transformed in our kitchens or by the food industry into the products we consume every day, and the complex principles of flavor chemistry, heat-induced reactions, microbiology, microstructure, and physics that give food products their characteristics such as taste and texture.

I have been a professional academic food scientist for almost 30 years and, in that time, my day-to-day experience of research has changed hugely. Part of this has been due to the changing nature of scientific publishing. I did my PhD in the 1990s, a time when resources taken for granted by today's young researchers, such as databases, open access, online journals, online reviewing, and even casual use of email, were unheard of (let alone smart phones, tablets, and laptops that didn't weigh a couple of kilograms). This feels like a confession of having learned how to chip messages onto stone tablets in an era lost in the mists of time, but the pace of advances in this area of technology, as with so many other areas of our everyday lives, has been simply astonishing.

I published my first paper in 1997, in a journal of which I later became an editor and have remained so for 13 years. I have since published around 250 papers, mostly first-authored by my MSc and PhD students, and coauthored by many colleagues in my own university (University College Cork, in Ireland), as well as other institutions, in Ireland as well as in countries including Denmark, Holland, Switzerland, Australia, Israel, France, the United States, and Spain. I have also published around another 50 review articles in journals or book chapters, and have completed one book on food science for people who like food but don't know about the science of this huge family of highly complex biological materials (*Molecules, Microbes and Meals: The Surprising Science of Food*).

As part of my academic work, I review (as do most researchers) probably 20–30 papers a year for other journals, as well as occasional grant proposals (for funding agencies in several countries), student PhD theses, and applications for positions or promotions.

I have also, for over a decade, been closely involved in the training of young researchers and postgraduate students in scientific writing and communication, running regular courses and workshops throughout Ireland and advising students on the importance of publication and other forms of communication (especially nonspecialist communication) for the dissemination of their work and for their future career.

These are all the influences that have led me to write this book, particularly the last mentioned, in terms of knowing the questions that inexperienced researchers have at the start of their career, and the things which they are somehow supposed to learn (perhaps by osmosis) but are rarely explicitly told. I then approached the topics in this book with the experiences (positive and negative) of over two decades as author, reviewer, and editor to create a 360-degree look at the world of scientific publishing, as well as all other forms of scientific communication beyond the peer-reviewed paper.

Over these two decades, the nature of my work has changed in other ways, though. In my early career, I was doing the research myself, and frequently donning a white coat (and occasionally wellies when going to farms to collect milk samples). As my career moved on, such requirements for specialized garbing decreased, and almost stopped completely.

In general, in the early stages of their career, researchers spend most of their time doing the practical aspects of research, and a small amount of time communicating it (through their first presentations or posters at conferences, first papers, theses, and dissertations) but, as their career matures, the relative proportions of these activities completely invert and flip.

Today, as a senior scientist and research leader, the vast majority of my time is spent communicating. I write (books, papers, emails, reports, proposals, reviews, presentations, and much more besides), read and edit, talk science formally (through lectures to students, presentations at conferences, seminars), and talk science informally (with my students, my researchers, colleagues, collaborators, and many others).

Today, my job is thus essentially completely composed of communication, verbal and written. There is very little I do on a day to day basis that does not depend on my communication skills and require me to have a very good command of such skills, whether practicing them myself or evaluating those of others.

I guess I also do some thinking, but I think in many ways that is now intrinsically tangled up in communication, as I think by writing and talking and discussing. Scientists are not like Winnie the Pooh, finding a "thotful spot" to which they retreat and decide "now I will have a good idea." Thinking in science, perhaps defined as the processing of information and the creative emergence of ideas, occurs in many cases as part of, and through the medium of, communication, as well as by things bubbling up in times when communication is not possible, like while driving, showering, sleeping, or having holidays. Active scientists are often not only scientists from 9 to 5 but also become so immersed in their subjects that it is never too far from their minds and thoughts, so that ideas will come at any time, often when least expected.[9]

My point here is essentially that communication skills are at the core of research and, to be a good and successful scientist (or researcher in any discipline) really requires someone to be a good communicator. Critically, science cannot contribute to society without communication.

---

[9] This can occasionally lead to a disconnect from conventional life, which has informed the cliché of the absent-minded scientist (which in my view, occasionally in the mirror, bears a grain of truth).

This is why I have spent a lot of my career trying to help the next generation of researchers to develop these skills and, more broadly, is what led me to write this book.

While my research has all essentially been within the sphere of food science, particularly the chemistry and processing of dairy products from milk to cheese, I have always been interested in the broader sphere of modern science and will draw on this, as well as the history of science (another long-held interest of mine), in providing examples and anecdotes. I also hope this book will appeal not just to researchers in the broadest possible sweep of scientific disciplines but also those not directly involved in research but interested in this key part of modern civilization, which is the transmission of important information about the world we live in and how it works: dispatches from the frontiers of knowledge.

{ 2 }

The History and Future of Scientific Publishing

Although mankind has wondered about its world, designed ways to learn more, and passed on the knowledge thus gained for millennia, the scientific paper as we know it today is a relatively recent construct. A key development in the 17th century was the introduction of shorter forms of communication than the books favored in preceding centuries, and before this time some famous scientific discoveries were published in rather unusual manner.

The modern scientific paper is characterized by its straightforward and unambiguous construction, in which the key objective is to present the work in as direct a way as possible. This was not always the case, though, as Galileo had to present his radical theories about what exactly orbited what in terms of the earth, moon, and sun in the somewhat obfuscated form of a dialogue between two philosophers and a layman, taking different sides of the argument, over the course of four days. This was published as a *Dialogue Concerning the Two Chief World Systems* in 1632, and led to him being placed under house arrest, while the work was placed on an intriguing sounding *Index of Forbidden Books*.

This led to him having to be more careful with subsequent publications that doubled down on his theories. A number of years ago, I was very pleased to receive as a gift a replica, in a nice red box, of a vellum reproduction of the publication by Galileo titled *Discourses and Mathematical Demonstrations Relating to Two New Sciences* (often shortened to *Two New Sciences*), first published in 1638.[1] This book, when first written, was smuggled out of Italy,[2] where it was deemed, based on the reception his earlier book had received, unsafe to publish, even though it too took the approach of framing its ideas within a philosophical debate between three men, who represented different stages of Galileo's own thinking. It was

---

[1] In a very modern twist, the box also contained the same book as a PDF file on a USB stick!

[2] While Monty Python claimed that "nobody expects the Spanish Inquisition," this presumably did not include those proposing completely heretical findings that went directly against the prevailing heliocentric (that the sun, and everything else, revolves around the earth) doctrine of the time. It should be noted that Galileo's work remained banned until 1758, and only in 1835 was church opposition to the heliocentric model dropped entirely.

eventually published in the Netherlands by a publishing house established by a man called Lodewijk Elzevier (a name adopted after minor modification by the academic publisher Elsevier when it was founded in 1880).

The first acknowledged academic journal was possibly the humanities-oriented French *Journal de Scavans* in 1665, followed (60 days later—something was clearly in the air that year, besides the Great Plague[3]) by the first journal devoted to science, *Philosophical Transactions of the Royal Society* (where the word "philosophy" related to "natural philosophy," the then-typical term for science or wisdom, the same sense from which comes the "Ph" in "PhD"). The latter was first published in 1665 by the Royal Society of London and is still published today (although split into two sister journals, differentiated as A, for mathematical, physical, and engineering sciences, and B, for biological sciences). The first issue of the journal included the first report of the red spot on Jupiter and, in its first years, its published papers included Isaac Newton's first major paper, on a theory of light and color, and early microscopic observations by Anton van Leeuwenhoek that revealed for the first time the hidden world around us.

The original goal of this journal was to achieve faster, more independent, and validated publication of the latest research results, compared to the self-publication that was the predominant model up to that time. Its impact can be judged by the reported observation of the biologist T. H. Huxley, 200 years after its founding, that, if all books printed were destroyed and only *Transactions* remained, humanity would still have a pretty good record of its intellectual progress in that time.

In its early centuries, the development of *Transactions* mapped the development of modern academic publishing, as the roles of editors became more formalized and less independent, and their own knowledge and somewhat variable expertise was supplemented in quality control by the introduction of what was essentially peer review. It is also noteworthy that *Transactions* apparently ran at a financial loss for the first 270 years of its existence, remaining viable only thanks to the support of the Royal Society for its key contribution to the dissemination of knowledge.

In 1800, the *Proceedings of the Royal Society* first appeared, initially somewhat cumbersomely called the *Abstracts of the Papers Printed in the Philosophical Transactions of the Royal Society of London*, with the snappier title being adopted in 1854.

The journal *Nature* was founded in 1869 by an astronomer and physicist named Sir Joseph Norman Lockyer, the codiscoverer of helium. *Nature* stated its mission as, first, placing before the public the grand results of scientific work and scientific discovery and, second, aiding scientists by giving early notice of advances in any branch of natural knowledge throughout the world. Today, *Nature* still publishes papers (very selectively!) in any branch of science, and has spun off a family

---

[3] Which forced Newton in the same year to leave London for the countryside, where he undertook much of the work on mathematics, mechanics, and physics for which he would become best known.

of specialized sister journals, such as *Nature Biotechnology* and *Nature Clinical Practice*.

Perhaps *Nature*'s greatest rival, *Science*, was founded in 1880, with support from, among others, Thomas Edison and Alexander Graham Bell, but was not initially a success and ceased publication in 1882. While it began again in 1883, it remained on uncertain ground and endured frequent financial crises until it became the official journal of the American Association for the Advancement of Science in 1900.

The forms in which some major scientific discoveries were reported are shown in Table 2.1. It can be seen that publication moved from books to journal articles as the preferred mode of communication in the 19th century, and that certain journals dominate the list (especially *Nature*). It is also noteworthy that, for many discoveries (especially in physics, but also biology and chemistry) German-language journals had a very high profile up to the early 20th century.

A further very notable factor in the evolution of scientific publishing, not immediately obvious from this Table, is the increasing number of names on papers describing major discoveries. Up to the 20th century, discoveries and their announcement were typically the work of single individuals, frequently in book form, while the 20th century saw increasingly common copublication by pairs or small groups of researchers (like Watson and Crick on the structure of DNA or Penzias and Wilson on the discovery of evidence of the Big Bang, for example), while the mean number of authors per paper rose steadily, to an estimated value of 5–6 at present.

In the 21st century, the occurrence of huge author lists (sometimes referred to as hyperauthorship) has expanded as rapidly as the lists themselves, reflecting the huge teams of researchers involved in international projects in fields where "big science" is the norm, like genomics or particle physics. An early record was set in 1993 by a paper in *New England Journal of Medicine* on an international randomized trial on strategies for treatment of heart attacks, which had over 900 authors, from 15 countries (the consortium was called GUSTO—a snappy contraction of their formal title of Global Utilization of Streptokinase and Tissue Plasminogen Activator for Occluded Coronary Arteries). Eight years later, the report of the initial sequence of the human genome in 2001 had 2,883 authors, while the current record seems to be held by a paper on the Higgs Boson in *Physical Review Letters* in 2015, which has 5,154 coauthors (requiring 24 pages of appendices to a 9-page paper to list them all).

Another change in scientific publishing over the last number of centuries has been changes in the language used, which will be explored more closely in chapter 4.

How Are Journals Ranked and Compared?

Today, there are estimated to be tens of thousands of scientific journals published, across a huge range of fields from astrophysics to zoology. Within each field, not all journals are the same, of course, and scientists need to know which are the best. How are journals ranked and compared?

TABLE 2.1 Where Some Major Scientific Discoveries Were First Published

Discovery	Year	Publication
Earth revolving around the sun (Nicholas Copernicus)	1543	Book titled *De revolutionibus orbium coelestium*
Laws of planetary motion (Johannes Kepler)	1609	Book titled *Astronomia Nova*
The circulation of blood around the body (William Harvey)	1628	Book titled *De Motu Cordis* (On the Motion of the Heart and Blood)
Earth revolving around the sun (Galileo)	1630s	Books taking the form of philosophical debates about the topics, presenting both the existing and (then radical) new theory
The existence of cells (Robert Hooke)	1665	Book titled *Micrographia, or Some Physiological Descriptions of Minute Bodies with Observations and Inquiries Thereupon* (published by the Royal Society of London)
Basic principles of mechanics, including the invention of calculus (Isaac Newton)	1687	Book titled *Philosophiae Naturalis Principia Mathematica*
That white light is composed of a spectrum of colors (Isaac Newton)	1704	Book titled *Opticks*
The fact that light is a form of electromagnetism (Michael Faraday)	1857	*Philosophical Transactions of the Royal Society*
The theory of evolution (Charles Darwin)	1859	Book titled *On the Origin of Species by means of Natural Selection, or the Preservation of Favoured Races in the Struggle for Life*
The discovery of X-rays (William Roentgen)	1896	*Nature* (in English)
The discovery of the electron (J. J. Thomson)	1897	*Philosophical Magazine*
The particulate nature of life (Albert Einstein)	1905	*Annalen der Physik*
The theory of special relativity (Albert Einstein)	1905	*Annalen der Physik*
The nucleus of the atom (Ernest Rutherford)	1911	*London, Edinburgh and Dublin Philosophical Magazine and Journal of Science*
Heisenberg's uncertainty theory	1927	*Zeitsschift für Physik*
The nature of the chemical bond (Linus Pauling)	1928	*Proceedings of the National Academy of Sciences*
The expansion of the universe (Erwin Hubble)	1929	*Proceedings of the National Academy of Sciences*
The discovery of penicillin (Alexander Fleming)	1929	*British Journal of Experimental Pathology*
The existence of the neutron (James Chadwick)	1932	*Nature*
The structure of DNA (James Watson and Frances Crick)	1953	*Nature*
The structure of proteins (Max Perutz)	1960	*Nature*

(continued)

TABLE 2.1 Continued

Discovery	Year	Publication
Proof of the Big Bang (Penzias and Wilson)	1965	*The Astrophysical Journal*
The genetic engineering of life (Jackson, Symons, and Berg)	1972	*Proceedings of the National Academy of Sciences*
Sequencing of DNA (of a virus, by Frederick Sanger)	1977	*Nature*
The viral cause of AIDS (Gallo and Montagnier, separate papers)	1983	*Science*
Hole in the ozone layer (Farman et al.)	1985	*Nature*
The existence of C60 (Buckminsterfullerene)	1985	*Nature*
Life (or maybe not) on Mars (McKay et al.)	1996	*Science*
First cloning of a mammal (Dolly the sheep, Wilmut et al.)	1997	*Nature*
The sequence of the human genome (International Human Genome Sequencing Consortium)	2001	*Nature*
The existence of the Higgs Boson	2012	*Physics Letters B*
The existence of gravitational waves	2016	*Physical Review Letters*

The main way in which this is done is that journals are ranked on various parameters by organizations such as the Institute of Scientific Information and Clarivate, who evaluate indices that quantify the quality and importance of the papers published in specific journals.

All approaches use a single common currency, the key countable element on which calculations are made, and this is the *citation*. Defining a citation is simple; when a researcher refers in their paper to another paper, whether as background, for a method they have used, or to explain or discuss their own findings, this is a citation. If a paper, in the years after publication, is never cited by anyone else, this is perhaps reasonably taken as an indication that it has not had a notable impact, while increasing numbers of citations are deemed proportionate to interest and impact.

Citation-based metrics used for the purpose of comparing and ranking journals include the following:

- Impact Factor: the ratio of the number of times articles from the journal published in a two-year window have been cited to the number of papers published in that journal in those two years;
- Immediacy Index: a measure of the speed with which articles in a journal are cited, such as the average number of times an article is cited in the year in which it is published;
- Cited half-life: median age of articles that were cited in the year in which the analysis is performed;

- Eigenfactor: like the Impact Factor, based on citations to articles in a journal, but more nuanced, in that citations from highly ranked journals influence the parameter more than those from less prestigious sources (similar to how Google's algorithms for ranking search results works). It is also influenced by the number of articles a journal publishes per year, and also takes into account disciplinary citation norms, ignores self-citation to articles in the same journal, and uses data over a 5-year window (compared to two years for the Impact Factor);
- SCImago Journal Rank (SJR): this parameter is calculated on similar principles to the Eigenfactor, but is based on data from the Scopus database rather than Web of Science, uses a 3-year rather than a 5-year window, and allows a limited level of self-citation by journals.

The top-ranked journals in July 2019 are listed in Table 2.2. Many interesting observations about the modern scientific literature are apparent from this table. First, the top-ranked journals are completely dominated by biology journals, and those concerning cancer research in particular (with the exception of two general journals, *Science* and *Nature*, and one physics review journal; the first pure chemistry journal, *Chemical Reviews*, is a review journal at sixth place).

TABLE 2.2 Top-Ranked Scientific Journals, as of July 2019, Arranged by Impact Factor

Rank	Journal Title	Impact Factor
1	*CA—Cancer Journal for Clinicians*	223.7
2	*Nature Reviews Materials*	74.5
3	*New England Journal of Medicine*	72.41
4	*Lancet*	59.1
5	*Nature Reviews Drug Discovery*	57.6
6	*Chemical Reviews*	54.3
7	*Nature Energy*	54.0
8	*Nature Reviews Cancer*	51.9
9	*Journal of the American Medical Association*	51.3
10	*Nature Reviews Immunology*	44.0
11	*Nature Reviews Genetics*	43.7
12	*Nature Reviews Molecular Cell Biology*	43.4
13	*Nature*	43.1
14	*Science*	41.0
15	*Chemical Society Reviews*	40.4
16	*Nature Materials*	38.9
17	*Reviews of Modern Physics*	38.3
18	*Cell*	38.2
19	*Lancet Oncology*	35.4
20	*Nature Reviews in Microbiology*	34.7

In recent years, it seems that more scientists are working in biology than in other broad subject areas, and hence citing papers in that field; large-scale initiatives such as the Human Genome Project and subsequent genomics projects certainly contributed to this dominance of biological research. While large-scale groupings of researchers are also involved in major expensive projects in physics, such as the massive (pun slightly intended) international collaborations focused on building blocks of matter such as the Higgs Boson, it may be that the number of other groups available to cite these papers is smaller than in biological research.

This is borne out by data from the analytical firm Clarivate in 2017 which suggested that the greatest numbers of citations were to papers in clinical medicine, chemistry, physics, biology and biochemistry, and molecular biology and genetics, which had accrued 35.9, 25.4, 13.2, 13.0, and 11.6 million citations, respectively.

The list in Table 2.2 also shows a high number of journals that publish review articles, which perhaps reflects the fact that such journals are favored in papers as a way to save authors from citing huge numbers of individual papers.

As in almost all aspects of life, technological advances in communication and information technology, particularly within the last 10–20 years, have revolutionized the scientific publication industry. Of particular relevance to the way in which scientists do their work is the development of a second (or increasingly sole) form of existence of scientific papers; as well as their paper (hard-copy) existence, on the physical pages of a printed journal, they have a double that exists on the Internet, housed in massive databases such as Science Direct and exchanged by email across the globe as PDF files, and searchable through yet more databases.

Not that long ago, undertaking a literature survey was a hugely time-consuming activity requiring access to a good library, and forcing researchers to pore over huge volumes of citations or microfiches of journal tables of contents. This has transformed in a relatively short time into a situation where researchers can, from their own office or home (or even phone), search through thousands of journals in seconds, and readily download copies of desired papers to their desktop directly, to be read or printed. The number of hours of labor saved by students and researchers thanks to this advance is simply incalculable.

One analogy is the way in which digital music players, or phones with this capability, have increasingly allowed music-lovers to transport the equivalent of entire vast record collections in their pockets, allowing instant, convenient, and user-friendly access to huge swathes of information, in one case papers, the other albums or songs. In both music and science, we have got used to key elements of information and usage no longer being physical entities, but electronic files, perhaps sacrificing direct physical interaction and tangibility for access and scale.[4]

---

[4] The analogy between digital copies of music and papers deepened with the emergence of a website called Sci-Hub, which was often described as being analogous to the site Napster, which was launched

The electronic revolution has also changed the physical nature of journals, with the idea of an "issue," which was a physical entity binding together a set of papers ready for publication at a certain time,[5] becoming increasingly meaningless, and online journals even eschewing volume, issue, and page numbers in favor of article identification codes such as Digital Object Identifiers (DOIs).

These changes have also had profound implications for researchers' relationship with the literature. In the days of hard printed copy, getting access to a paper meant going to a library and making a photocopy, time and effort which has obviously been greatly reduced by online access. However, those who remember such days will invariably also remember the experience of finding unsought other papers, which perhaps happened to immediately precede or follow the one in question, and which turned out to be highly interesting and possibly sparked off another idea or research direction. Targeted searches through databases reduce the opportunities for such unexpected findings. So, researchers should try and keep an eye on the broader literature in different ways, such as through subscribing to email notices of new papers in their areas or tables of contents of new issues of journals in their area. In parallel, the deluge of material that can result when undertaking a database search has led to the recognition of a new set of competencies which a researcher needs to master, called *information literacy*, or the skill of being able to tailor search strategies of the literature as effectively as possible.

The Significance of Impact Factor

The primacy of the Impact Factor as a means of judging the "quality" of a scientific journal has been questioned in many ways. One problem is the potential for the parameter to be "gamed" by journals artificially manipulating their own Factor, for example, by encouraging authors of papers they are about to publish to cite more papers from the relevant time window in that journal. There have been quite a few cases where this has been detected, and journal editors have been removed from their posts for such behavior, or in some cases journals "delisted" from the Impact Factor list for such practices.

---

in 1999 to make huge amounts of music available for sharing between individuals outside copyright restrictions. Sci-Hub was launched in 2011 as a means of subverting the paywalls behind which so many papers were secured, by similarly allowing peer-to-peer sharing of files, this time of papers rather than songs or albums.

[5] Although the wonderful Feedback section of *New Scientist* magazine noted in April 2019 that the journal *Pure and Applied Mathematics Quarterly* had just published a 600-page issue containing just five (long!) articles by just one author, the physicist Ed Witten.

Any search of Impact Factors across fields shows that they are also very field-specific. For example, a quick search of the range of the top five journal Impact Factors in a range of disciplines in Web of Science InCites data shows the following spread of values:

Astronomy and Astrophysics	9.3–24.9
Computer Science (Artificial Intelligence)	8.1–11.5
Dentistry and Oral Surgery	4.1–6.2
Food Science and Technology	5.2–9.5
Horticulture	1.9–3.9
Oncology	24.6–244.6
Ornithology	1.9–2.7
Zoology	3.6–6.8

So, a good Impact Factor in one field is not the same as a good Impact Factor in another. The journal I edit (*International Dairy Journal*) has an average Impact Factor of around 2.50, which is quite respectable within the field of dairy (or food) science, would be very high for ornithology, and is barely notable for cancer biology. Researchers considering journals to which they should submit their papers should consider the appropriate scale against which they will be judged, in terms of where the best place they could reasonably be expected to have published, rather than the massive scale of all possible Impact Factors.

The InCites data can also rank disciplines by median Impact Factor, on which basis the top five in 2018 were Cell and Tissue Engineering, Allergy, Cell Biology, Oncology, and Immunology (followed by Rheumatology, Gastroenterology and Hepatology, and Endocrinology and Metabolism). Medicine and Biochemistry once again dominate any analysis of citations, but I guess it is reassuring to see that more scientists are working on keeping us alive and well than anything else!

In recognition of ongoing discussions of the use and interpretation of Impact Factors, a meeting of the American Society for Cell Biology developed in 2012 what is known as the San Francisco Declaration on Research Assessment, which articulated the problems with Impact Factors, including what has already been discussed here and also the facts that they are very skewed in any field, are highly field-dependent, and are calculated on the basis of data that are not transparent or openly available.

The Declaration recommends, for these reasons, that Impact Factor data not be used for decisions on funding, appointment, or promotion, and that research quality be judged on the basis of individual articles rather than the journal in which they appeared.[6] Such concerns have also increased discussion of the use of alternative metrics, such as the Eigenfactor and SCImago mentioned earlier.

---

[6] Although this remains common practice today.

TABLE 2.3 Top-Ranked Scientific Journals, as of June 2019, Arranged by Alternative Metrics

Journal Title	Eigenfactor Ranking	Total Citations Ranking	SCImago Journal Ranking
PLoS One	1	4	
Nature	2	1	
Nature Communications	3		
Science	4	2	
Scientific Reports	5		
Proceedings of the National Academy of the USA		3	
Journal of the American Chemical Society		5	
CA—Cancer Journal for Clinicians			1
Morbidity and Mortality Weekly Report			2
Nature Reviews Materials			3
Quarterly Journals of Economics			4
Nature Reviews Genetics			5

In Table 2.3, the top journals ranked by some of these alternative metrics in 2019 are presented and, while each bears some resemblance to the Impact Factor rankings, in each case some journals that do not appear in Table 2.2 appear.

Selecting a Journal for a Paper

One of the key decisions to be taken once a piece of research has reached the point at which it is ready to be written up and submitted to a journal is the journal to which it will be submitted.

For most manuscripts, the authors could readily prepare a list of journals to which their work could be submitted (reflecting, among other factors, the journals that have already published work in the field, and hence appear in their reference list). On this basis, a list of 5 or 10 or more journals that could possibly consider their work could probably readily be drafted, but what criteria will then be applied to decide which to send the work to?

One suggestion that might seem plausible to an inexperienced researcher is, if say five journals could be listed as possibilities, to send the paper to all five journals at once, and let the first to accept have the "honor" of publishing the paper. Needless to say, contemplating publishing the same paper in more than one journal is premeditation of serious plagiarism, but also submitting to multiple journals, even without the intention to eventually publish in more than one of them, is not acceptable practice. The reason for this is that shepherding a paper from manuscript to published article involves a huge amount of effort, including by reviewers who are doing their part on a voluntary basis. Allowing this effort to be expended and then turning around and saying that it was wasted because the authors have received a better offer from a different journal is simply professional discourtesy of the most serious order, and is prohibited by most modern submission rules.

So, all-at-once submission is not an option, but the journals can still be ranked in order on the basis that, if the first does not accept the paper, then the second can be tried and so forth (multiple journals can be approached with the same, or similar, manuscript, only in sequence, not in parallel). The actions taken when a manuscript is rejected will be discussed in detail in chapter 6.

But in what order is this list ranked? Some key criteria might be as follows.

JOURNAL RANKING

One of the most important ways in which a paper's quality will ultimately be judged is the prestige of the journal in which it appeared. On this basis, it might seem logical to rank the journals in Impact Factor order (or perhaps using one of the other metrics mentioned earlier), and start at the top. In fact, following this principle to its logical conclusion, a general science journal like *Nature* or *Science*, or both, which publish papers in any field of scientific research, would appear at the top of every such list.

However, in practice, most researchers will only consider these journals for exceptional pieces of work. They know the standards are so high, and the process of peer review so tough, that sending papers that are not of that level is only wasting the researchers' time and the journals' effort, while also inviting an unnecessary and probably inevitable rejection and blow to their confidence, which is something noone should rashly encourage.

So, taking these essentially inaccessible journals off the list, there is still a need to make a similar judgment between the possible journals that remain. In other words, one of the hardest questions for a researcher to ask themselves, particularly at the start of their publishing career, is "how good is this work actually?"

This basic but hugely important question tests whether the researchers believe the work is of sufficiently high standard to be accepted in the highest-ranked journal on their list, or whether it is more realistic to aim for a lower-ranked journal. There are no easy answers to this question, but making such judgments becomes easier as a researcher gains experience of publishing (based on their history of scars earned in the publishing school of hard knocks), and learns to calibrate their expectations and ambitions based on whether certain previous papers were or were not accepted by specific journals. New manuscripts can then be intuitively targeted at one or other journal based on this experience.

Researchers will have papers they are very proud of, that they believe will get into the very best journals that could be considered. On the other hand, researchers will also sometimes have pieces of work which they know is not at that level, or quite robust enough (for example, in terms of replication) but which they believe will still have an interested audience and would like to make available. Every field typically has such journals that offer a home, perhaps after lighter peer-review, to such work. Researchers will know that too many publications in such journals might not make their CV look too impressive, but a low level of publication there will not reflect badly. It has also been suggested that some researchers might deliberately target "easier" journals to publish work that might be controversial for more mainstream journals.[7]

Of course, past performance is no guarantee of future success, and there will always be journals that appear on a list with which the researchers involved have no history, but such careful considerations can greatly increase the likelihood of a rapid and successful outcome on submitting a paper.

A reasonable question might concern a journal that is new not just to the authors but to the field, and how to place this on such a list. Every week, new journals are appearing, some fit only for utter avoidance, as we will discuss later, but others launched by reputed publishers. A quick check of the Elsevier website suggests that this one publisher alone launched at least 60 new journals in 2017. The most common indicator of journal standing, the Impact Factor, is of no help in such cases, as a journal is not assigned an impact factor for 5 years after it first appears, as obviously time needs to pass for the journal to accumulate a body of published articles, and citations to those articles.

So, how can an infant journal less than 5 years old be judged in terms of clues as to its likely ultimate reputation, so that authors can entrust it with their precious paper? Key indicators would include the following:

Who is publishing it? If the journal has a major publishing house behind it, or known and respected professional society, then that is a major vote of confidence. If the ownership is unclear, or decidedly dodgy, then red flags should start to appear and wave frantically. An important step could also be googling the publisher's name with the terms "reputation," "controversy," or "predatory publisher." It should be noted that some large publishers, such as Elsevier, have been the subject of campaigns to boycott them based on concerns over their business practices, which may be another reason that researchers may decide to look carefully at a journal's publisher.[8]

Who is managing it? Each journal website should list those responsible for the scientific standards of the journal, specifically the editors but also the editorial board members. If we assume these are honest and true lists,[9] they provide an insight into the standing of the journal in terms of the reputation and standing of those who have agreed to invest their time (and name) in it. A bit of googling of unfamiliar names should give a quick impression of their seniority and credibility, and further conclusions can be drawn from the standing of the institutions in which these individuals are working, as well as the international (or otherwise) mix of those named.

---

[7] "Predatory publishers: the journals that churn out fake science" *The Guardian*, August 13, 2018.

[8] In the interest of full disclosure, the journal of which I have been an editor for over 15 years is an Elsevier journal.

[9] There have been (unfortunately) cases where respected researchers have had their names used in this context without their permission.

Who is publishing in it? Assuming this is not volume 1, issue 1, of the journal (or online equivalent) an absolutely critical indication of quality will be the authors of the papers published to date. As for the editors and editorial board members, key questions will revolve around authors' previous history, reputation, and location. If good people in a field are taking a chance on this journal, that makes the path less hazardous to follow.

How professional does it look? For many disreputable journals, it is unfortunately very easy to make judgments based on the appearance of a website, and the professionalism with which a journal portrays itself to the world. A researcher's gut instinct will usually detect a journal of questionable standards (such as typo-riddled text, cheap looking page design, poor English expression, overuse of exclamation marks, or perhaps any use at all of these).

MATCH TO TOPIC AND AUDIENCE

When I started publishing papers, a key consideration was the fit of the topic to the area covered by a specific journal. Twenty or so years ago, the rationale for this was simple. For a paper to "succeed," in terms of having an impact or benefitting the research, reputation or career of a researcher, the key first step was for it to be read, and for it to be read it must first be found. In the antediluvian predatabase days, when journals appeared only or predominantly in physical hard copy, this meant that other researchers who would read (and then make use of) the paper needed to have a subscription, either institutional or personal, to the journal, and be likely to browse it frequently. So, careful selection of the right journal was critical in ensuring this match-up of reader and paper.

If a paper was published in a journal that was not frequently read by the "right people," then it could be essentially overlooked, or depend on more circuitous routes to meet its audience, such as tracking through later citations in other papers or through huge printed books of abstracts or contents of journals (I remember peering and squinting as a PhD student at tiny font in a massive tome called the Science Citation Index, a voluminous hard-copy precursor to today's Web of Science). Frequently, when such methods did identify a paper of interest, it was not to be found in the searcher's library, and a postal request called an Inter-Library Loan form or suchlike had to be dispatched, hopefully resulting in the delivery of a copy of the paper several weeks hence.

There is no doubt that modern database systems have reduced the time required to undertake a significant review of the scientific literature on a topic by several orders of magnitude, but this is not without cost, both literal (as libraries need to pay significant fees to maintain journal access) and more related to how research is done. As mentioned earlier, researchers who came of age before the Internet found many articles that led to new ideas not because they were looking for them, but because they found them while looking for something else in the same journal issue, while the newer generation of researchers has to develop skills of information literacy to help them filter through the information overload that can result when they undertake a search.

It could be argued that, in the modern publishing landscape, papers getting "lost" is far less likely to happen as, once papers are published almost anywhere, the Internet and databases will find them. Then, if the person who finds them does not have an institutional subscription to that journals' articles, they can pay an access fee or simply find an email for the corresponding author and request a copy.

So, does this mean that where the paper is published has become less critical?

The answer is yes only in the specific context of access to the article and the ease with which the target audience can find it.

The choice of journal still conveys a lot to the reader in terms of judgments they will make about the paper, such as the impact factor and standing of the journal, for one thing. Readers know that higher-level journals will have subjected the work to intensive peer review and scrutiny, and hence assume that papers appearing in such journals come with a stronger stamp of quality control, as it were.

In addition, there is a question of "marketing" the paper that is implicit in the selection of the journal to which it has been submitted. Who do the authors think are the audience for this paper? What does it say about their impressions of their own work that have caused them to select this journal? Do they see it as international or more local interest? In particular, the choice of journal is particularly significant for interdisciplinary work, such as a paper on medical microbiology that could be sent either to a clinical/medical journal or a microbiology journal. In such a case, what do the authors think is the key thrust of the paper and who do they think are the key audience for the outcomes of their paper (and thus what do they think are the main outcomes)?

The selection of a journal thus less directly influences its audience, but yet in other ways it greatly impacts on what the audience think about the work, and even how the audience thinks the authors perceive their own work.

Finally, every journal will somewhere have articulated a definition of the areas of research in which they will consider papers; this may be called "Aims and Scope" or something similar. In every case, authors should check these details against the topic of a paper they are thinking of submitting, to ensure that the paper is definitely in scope, as otherwise the editor is likely to reject it without review. Such a rejection is not a judgment on the scientific quality of the paper, and the authors can simply resubmit it elsewhere, but this will have cost time and effort (for the authors in formatting the paper for the first journal, and the editor for reading and coming to a decision, however rapid, on it) and it simply looks sloppy.

EDITORIAL POLICY

Experienced researchers who have, as mentioned earlier, accumulated a certain amount of scar tissue through their interactions with journals on previous papers will apply other factors to the selection of a journal. These inevitably relate to human factors and personal/professional interactions.

Of course, these should have no bearing on the acceptability of a paper by a journal but, as with so many aspects of real life, human factors count for a lot more than might be ideal. For example, how have the authors of a paper previously interacted with a journal to which they are submitting a paper? Have they had papers regularly accepted, regularly rejected, in that journal? Are they a member of the editorial board or have they often reviewed for them? Do they know the editor? All of these factors may weight positively or negatively for an author, as prior relationships with a journal could mean that the most rigorous review possible can (and possibly should) be applied to ensure no appearance of favoritism or a "fast track." It may seem inappropriate for an editor of a journal to be coauthor of a paper submitted to that journal, but if it is a significant journal in that field it may be unfair to their students or collaborators to exclude the possibility of their publishing there on this basis, and once again the journal must ensure that no unfair advantage ensues.

In terms of personal relationships with editors, I once had a student who, after a particularly harsh treatment by a journal editor (who was notoriously cantankerous), drafted an email to me with rather choice language about the situation, and sent it to the editor by mistake rather than me. As a result of this, we avoided sending any more papers to that journal until the editor retired (not too long afterward). Human relationships matter!

Some journals practice blind peer-review (to be discussed in chapter 6), where the referees evaluate submitted manuscripts without any indication of who the authors are and where they come from. This removes one significant element of the human side of science from a critical part of the publication process and should ensure that all authors are treated equally, free of biases, grudges, or positive or negative connotations created by their name or affiliation.

In another development, the Centre for Open Science operates what are called Registered Reports, where peer review is applied before a study is even complete, in an effort to prevent the publication of poorly designed studies. In this model, researchers submit a manuscript outlining their proposed approach and methods, and possibly pilot data, and these are reviewed; if deemed of sufficient quality, "In principle acceptance" is noted, and when the final paper is submitted a second round of peer review is applied to determine final acceptance.

The majority of journals do not practice such measures, however, and it will be interesting to see in coming years whether this practice becomes more widespread.

FINANCIAL AND PRACTICAL CONSIDERATIONS

There are a number of other considerations that may influence the selection of a journal for a particular paper, including publication charges (can the project budget pay for charges, whether for open access or for a "traditional journal" that levies page charges?), frequency of publication and speed of publishing, page or word count limits imposed, and more.

Someone once said that one should consider whether one would publish a paper if the devil offered you a large sum of money to tear it up and forget about it; if one even hesitates, the paper is not worth publishing. Today's authors face a somewhat similar dilemma—is a particular journal worth the cost of the paper appearing there when alternatives (not necessarily of lower impact or repute) that publish for free are available? Today, the issue of page charges also arises in the context of open-access journals, to be discussed in what follows.

In addition, practical considerations might arise due to local preferences and pressures in the authors' institution(s). There might be expectations that researchers publish in certain journals, or journals with a stated minimum standing in their field, while institutions or funders may also insist on open access publication. In some cases, such pressures might even take precedence over other factors mentioned earlier.

Models of Publishing

One of the most significant developments in scientific publishing in recent years has unquestionably been the emergence of what is called open-access publishing.

In the traditional model of publishing, scientific journals are published by one of two main models:

1. A large commercial publisher (such as Elsevier, Springer, or Wiley) publishes journals in which authors can publish papers without charge. However, for readers to access the articles, they (or their institution or employer) must have a paid subscription to that journal, or they must pay a per-article charge to access the article;
2. A smaller professional organization, such as a scientific society, publishes a journal but, to cover their costs (editorial office, printing costs, etc.) charges authors a charge per published page of finished articles (in my experience, this charge might be in the region of $100 per page), as well as charging subscribers later for access to the journal or individual articles.

In the last few decades, a revolution has taken place through introduction of a third model, called open-access publishing. In open access, there are no barriers in the form of required subscriptions or per-article charges, and anyone anywhere, with suitable Internet access, should be able to find and read the papers and benefit from their contents. Open access, as the name suggests, is meant to truly democratize the possibility of use of scientific information.

Recent years have also seen the emergence of several hybrid models of the first two models with the open-access model, where articles are published by large commercial publishers in a manner that entails no or fewer restrictions on being able to access or use them. For example, authors may pay to publish their articles open-access, or publishers may allow articles to be offered open-access after a certain time period or under specific circumstances.

Before discussing these models further, it is necessary to consider the significance of the concept of copyright, which underpins these developments.

The Copyright Conundrum and
the Concept of Open Access

A key driver behind the open-access model is the copyright implications of the
original two models. In both of these, a step that happens just before the acceptance of a paper that has been deemed suitable for publication following thorough
peer review is the signing of a copyright transfer form. Signing such a form, electronically or physically, means that the authors of a paper assign the rights to the
published version of their paper to the publisher.

It should be noted that this refers very specifically to the published paper exactly
as it appears in the journal, with journal formatting, logos, and specific layout. The
authors retain rights to the text that was used to generate this version.

So, picture two documents side by side, one a long word-processed document in
manuscript format which has been accepted for its scientific content and quality by
the journal, and the other a PDF formatted version of the same material exactly as
it would appear on the journal website (or hard copy, if such existed).

Now, picture a researcher proud of this paper, and eager to share the output
and show off their achievement. Which can they use? If they put the former,
manuscript-style, version of the paper on their homepage, university department
website, or perhaps LinkedIn page, there is no problem, but if they instead choose
to upload the PDF they have in fact broken the law, by infringing the copyright of
the publisher.

If, at a later stage, another researcher read the paper, and decided they would
like to use an image from that published paper in their own work, with appropriate acknowledgment of source, they need to seek permission to do so. While
intuitively it may seem logical that they would seek such permission from the
authors of the paper, it is actually the publishers of the journal in which the
paper appeared who have to grant this permission (and normally do so without argument, once a formal request process is followed, usually easily accessible online through the relevant journal website), because they now hold the
copyright!

Not only that, but, if the original authors decide at a later stage to include a key
figure or table from their paper in a later publication, say a review article where it
makes sense to include this level of detail, they even need to request permission
from the publishers to do so.

So, to summarize, researchers do their work, write up papers, and submit
them to journals where, if peer-review recommends publication, the work finally
appears in a format created by the journal. That formatted article is now essentially the property of the publisher, and they legally can control all access to that
article, normally through the existence of a paywall, which means that only paying

subscribers to the journal or those willing to pay a per-article charge can see the paper file.

One way this traditional scenario has been described is that the work of scientists is essentially taken and used by commercial publishers, frequently large corporations, who benefit from the unpaid work of not only the authors but also a huge number of scientists undertaking peer review, as well as that of editors paid a small honorarium for their time. The companies then make profit by charging everyone for access and strictly controlling what can be done with the published work (even by the authors of the individual published articles themselves).

To those without journal subscriptions, or the means to pay for individual articles, the work is inaccessible and cut off. This could lead to situations where, for example, reports of research describing medical advances relevant to diseases in some of the poorest countries on earth are completely unavailable to doctors, patients, or researchers in those countries for whom the impact of this research should be greatest, or journalists seeking to write about it.

The Open-Access Uprising

It was exactly these types of concerns, relating to the requirement of researchers to sign over critical rights to their own work to companies who then proceeded to make money from this work (and the fact that a huge amount of scientific information was then controlled in terms of who could access and benefit from it[10]), that led some key researchers to call for a whole new approach to the publication model.

The principle of open access that has emerged in response to these concerns is based on the principles that, first, authors should retain all rights and ownership of their work and, second, access to that work should be free and open to all, with every appropriate opportunity for others to use or reuse the information therein at a later stage, once the source and authors are correctly attributed.

Copyright in this model remains with the authors. However, as the publisher cannot recoup costs of publication through sale of access rights and subscriptions, such costs must be borne by the authors.

The first open-access journals appeared in the 1990s, and the number of journals has grown rapidly since then, leading to 4,769 journals publishing 191,000 articles in 2009,[11] a growth rate of 30% compared to 3.5% for "traditional" publishing in this period. There has been a further huge rate of growth in the decade since that analysis; an online directory of open access journals (doaj.org) counts 13,586 "high quality . . . peer reviewed" journals, and 4,154,556 articles. Today, in

---

[10] This argument is supported by reports that 65 of the 100 most cited papers in the scientific literature are paywalled, with an average cost for access of just over $30; the cumulative profit made from this access is enormous.

[11] Laasko, M., Welling, P., Bukvova, H., Nyman, L., Björk, B.-J., and Hedlund, T. (2011) The development of open access journal publishing from 1993 to 2009. *PLoS One*, 6, e20961.

many fields there are very reputable open-access journals with respected editors and high Impact Factors, but there is also, as discussed in the next section, a large number of far from reputable publishers and journals in this space.

A number of different models of open access are recognized today. "Gold" open access refers to publication such that the publishers offer open access to the publication without restriction, with the authors paying the costs of publishing their work, while "green" open access refers to the depositing of articles some-where else (usually not the published typeset journal version, but the authors' final manuscript), usually in open-access repositories, where they can be accessed for free.

As this area evolves, further models are emerging, such as cases where article charges are not paid by the authors, but may be covered by an institution, govern-ment, or philanthropic body, or avoided due to the publishing work being done by volunteers ("platinum" or "diamond" open access).

Of course, the emergence of open access represented a threat to the business model of the commercial publishers, and their practices had to evolve as a result. One hybrid model allows authors to select to have their paper in a "traditional" journal published open-access, without restrictions, and tables of contents for journal issues today increasingly show a badge or label of open access on certain papers, where the authors have (critically) paid for this option. Such costs are typi-cally of the order of several thousand dollars, a not-inconsiderable sum, and much more than even page charges in non-open-access journals would be. In other cases, a sort of rolling paywall may operate, with papers becoming freely available after a certain period of restriction has passed.

The principle of open science is every year being more and more embedded in scientific activity worldwide, with goals such as those expressed by the European Commission of making research "more open, global, collaborative, creative and closer to society." This involves more than just free access to published papers, but also increasingly open sharing of experimental data and research outcomes. In an illustration of this approach, the European Open Science Cloud seeks to offer researchers a virtual environment in which to store, share and reuse large volumes of research data.

In 2018, a group of 12 European funding agencies formed a consortium called cOAlition S (OA standing for Open Access and S for science, speed, solution, and shock), with a stated goal of making all the research they publish open-access (such that, as their website says, it "cannot be monetised in any way") by January 1, 2020. This is clearly a very ambitious goal, but is underwritten by the fact that these agencies between them control 18 billion euros worth of research funding, and several more agencies have signaled their intent to join this coalition of the willing. cOAlition S also plans to provide incentives to establish open-access jour-nals and platforms in areas where high-quality ones do not yet exist, does not accept hybrid publishing (where commercial publishers allow researchers to pub-lish individual articles open-access for high fees) as being consistent with their principles, and rather ominously states that the funders involved will "sanction noncompliance."

In the United States, the National Institutes for Health mandated in 2008 a policy that all papers resulting from research that it funded should be made available for free online within 12 months of publication, while an initiative called the Open Research Funders Group, which comprises multiple funding agencies, including the Bill and Melinda Gates Foundation, seeks to accelerate access to research findings and data.

It is clear that the outcomes of research, from raw data to finished articles, will in the future be far more freely available, taking advantage of the online world, than ever before. This is a rapidly changing field, in which organic change led by researchers has been reinforced by some major players lending their support, money, and pressure, and it seems inevitable that the landscape of scientific publishing in 5 years will be very different to that found today.

As well as benefits to the world and readers, open access has obvious advantages for the researchers who publish in this way, such as enhanced impact and appreciation of their work, and greater visibility and hence reputational benefit. There are also reported to be citation bonuses for open access papers (particularly in the short term after publication), as more readers inevitably means more potential for those readers to cite these papers in their own work.[12]

Are there any downsides to open access though?

The Problematic Side of Open Access

The goals of open access, as articulated in the previous section, are unquestionably laudable and fair. However, when it comes to the implications for publishing, significant issues have emerged in recent years.

If we strip back the principles of open access to a simplistic model, a researcher submits work to a journal and (following, presumably, peer review) the journal charges them a fee to make their work available, in almost all cases, only online.

Let us imagine a hypothetical scenario where someone takes a paper, pretends to peer-review it, and then puts it up on a website that has the facade of a scientific journal after charging the authors a handsome fee for the privilege.

Could this actually happen? Surely not?

Unfortunately, this is a scenario that is astonishingly, and infuriatingly, far from hypothetical today, and recent years have seen the emergence of shockingly high

---

[12] The *Nature* website reports that the Wellcome Trust reports that open-access papers arising from research they funded were downloaded 89% more than access-controlled papers.

numbers of completely untrustworthy, uncredible agents purporting to be scientific journals but which are actually essentially a borderline criminal activity.

The term for publishers who set up such journals is "predatory publishers," language that is appropriately loaded, with connotations of monsters seeking to feast on or take advantage of others. In essence, what such publishers do is seek to take advantage of the work of scientists and provide what appears to be a serious outlet for their work, but which is in fact far from this. An American university librarian, Jeffrey Beall, is credited with coining the phrase "predatory publisher" and has been a key chronicler of the rise of untrustworthy publishers and journals, of which he has kept a rapidly growing log that can be easily found online (although the regular target of complaints and lawsuits by publishers listed). It has been estimated that "predatory open-access publishers" had published 175,000 articles[13] by 2018.

Researchers today are bombarded daily with email invitations to submit papers to (or even join the editorial boards, or become editors, of) journals with superficially legitimate titles for the field concerned but which are actually not fit for anything other than deletion after a moment's glance.

In my own experience, certain things mark out such a suspicious invitation, including some or all of the following:

1. A sending email address from a general email service, such as Gmail or Yahoo, an indication that all further business such as submission will be handled by email, or a sender name that sounds vaguely impressive (email from Professor X, where Professor X is not found by any Google search or even listed on the journal website[14]);
2. The word "Greetings" in the salutations and a generally obsequious tone;
3. The excessive use of exclamation marks (e.g., "Greetings!!!!"[15]);
4. A high incidence of typographical errors, such as the journal whose name included the mysterious term "Scince," which was repeated several times in the email;[16]
5. A mistake in the name to which the email is sent (like surname and first name reversed, or the email being addressed to a coauthor different to the person who receives it);
6. A sense of urgency, particularly based on the journal needing just one more article by a certain date to fill an issue ("If you have ready manuscript in any topic related to the field of . . . please attach the document to this email as soon as possible for further process");

---

[13] "Predatory publishers: the journals that churn out fake science" *The Guardian*, August 13, 2018.

[14] One email that followed this trend led me to a website where the editor-in-chief boasted of having "published more than 20 papers in reputed journals, also with impact factor."

[15] In my view, a suspicious number of exclamation marks in a scientific paper or formal communication is any number above zero.

[16] Ticking off points 2, 3, and 4 in one go, one email I received started, verbatim, with the words: "Greetings for your successful life! We wish you a wealthy and Peaceful for this year 2018" [*sic*].

7. A dogged persistence in chasing authors who ignore the emails first time around ("This is just a friendly reminder . . . ");
8. Reference to questionable Impact Factors and indexing services that are not standard bibliographic norms;
9. A remarkably broad scope of areas of interest in which papers will be considered;
10. A promise of very rapid peer review (often 24 or 48 hours) and time to publication;
11. A promise of reduced publishing charges (but charges nonetheless . . .).

All these journals claim to operate peer review, yet I rarely get an invitation to review for them, although they seem eager to publish my work. Why is this? This is because, in most cases, the claim to their application of peer review is, as with all aspects of these "journals" a sham. It is noteworthy that traditional credible journals do not have to advertise their existence.

Various individuals have tested this claim in inventive ways in recent years. One group of researchers submitted a paper titled "Get me off your f***ing mailing list" to a journal called *International Journal of Advanced Computer Technology*. The paper text consisted of the above seven words repeated over and over, sometimes as headings, mostly as paragraphs of text, and even creatively in two figures.[17] Apparently, on submitting this paper to said journal, it was accepted (and deemed excellent by the, one can only assume entirely fictional, reviewers) and only when presented with a bill for the cost of publication did the authors withdraw and publicize the sham.

There was also an early project called SCIGen in the mid-2000s, where a computer program randomly generated computer science papers, which were then submitted to real journals, of apparently dubious standards, to test their willingness to publish complete gibberish. Some examples can be found online, and I particularly like the appearance of graphs with axes labeled things like "Time since 1977 (in teraflops)" and "latency (celsius)." Apparently, several such papers were submitted to conferences and accepted, leading to presenters volunteering to present them, on the basis that they do so having never seen the (computer-generated) slides in advance of stepping on stage, and wearing outrageous disguises while they did so.

In a more recent example, in 2013 a journalist from *Science* called John Bohannon used a computer program to generate a random selection of meaningless papers in the area of medicine and sent these to around 300 open-access journals. The papers were generated by a computer program based on the template of "Molecule X from lichen species Y inhibits the growth of cancer cell Z," where X, Y, and Z were inserted from a random database of entries. Each paper had identical (meaningless) conclusions that were clearly not compatible with the fake data presented, and had lists of fictitious authors. Just over half the journals

---

[17] The paper is readily findable online through Google.

approached accepted the paper, and a third rejected it (the rest did not respond by the time the article was written, or had shut up shop already). Journals from publishers on Beall's list accounted for over 80% of the acceptances.

Overall, the incidence of predatory or simply rubbish (non)journals is rising in all fields. For example, in my own area of food science I kept a log of invitations from such journals in the period January to July 2018 and, in that period, I received invitations to submit papers from 39 journals with topics relating to food, nutrition, or agriculture that I would regard as decidedly suspicious. All these journals had titles that seemed reasonably plausible, with some rather grand titles and a fondness for including "Annals," "Archives," or "International" in their names.

Of those, nine asked me to become an editor or join an editorial board, and around a quarter emailed me repeatedly, apparently being quite concerned that I had somehow failed to respond to their initial entreaty. Most gave a deadline by which they expected me to make my submission (to make their next issue) but only a handful mentioned peer review (and then it was typically double-blind peer review within 15 days). Thirteen different publishers were mentioned,[18] none with apparent academic standing, and many journals mentioned no publishers at all. The salutations of quite a few were aimed to former coauthors of mine, and some just started "Dear Dr" or "Dear Professor," while a few mentioned one or other paper I had previously published as being the reason for the approach (one mentioning specifically my personal favorite paper, which it listed as " ," i.e., a blank space). One flattered me by seeking the inputs of "one of the most brilliant and gifted minds of this century" (I bet they say that to all the authors), but got my name wrong, while one failed to spell the word "Veterinary" correctly in their own title.

This stuff, all in all, is reasonably funny and easy to dismiss, yet it represents drains on time and energy, creates noise and distraction in the system, and is a blight on the serious business of academic publishing. It simply should not be happening!

I also have no doubt that the incidence of this type of harassment is far higher in fields other than my own. I published some time ago a very short communication (effectively a letter to the editor commenting on a published paper) in a medical journal in which I have never otherwise published, and which is quite far outside my area of activity. This couple of hundred words did, nonetheless, catalyze an incessant deluge of invitations to submit papers and present at conferences in, or submit papers to, that medical field. In the period mentioned earlier, I received even more from journals with more medically oriented titles than for ones in food-related fields, and quite a few relating to physics or engineering as well, so I can only imagine that the mailboxes of those in that field for all their research are under constant bombardment by this turgid torrent of trash.

---

[18] Of which nine are on Beall's list, two are not but looked among the most suspicious of them all (so perhaps very new) and two are listed as being somewhat borderline, or moving toward greater respectability.

This is basically what worries me most about this type of SPAM (Scientific Papers? Absolute Mockery), which is that it generates a fog of obfuscation and doubt around what should be a highly professional and credible enterprise. It is relatively easy, if absolutely infuriating, for experienced researchers to sift through the swamp and ignore the noise but, for those starting out, the need to develop crap-filters and not end up chasing ephemera and dead-ends is ever more urgent, and this should not be the way things work.

Anyway, on that pessimistic note, we end this chapter, and next turn our attention to the structure and objectives of a scientific paper and how it has evolved a particular form, almost universal across disciplines, to transmit scientific information as effectively and credibly as possible.

What Does a Scientific Paper Say?

Having navigated the hugely complex landscape of the modern scientific litera-ture, staying to the trusted paths and roadways and avoiding the marshes and dark alleys of predatory journals that tempt the unwary traveler, researchers should come to a point where they have identified the best journal for their paper, and must now focus on writing the actual paper. We will explore this key part of com-munication on two fronts: the formal structuring of the paper in this chapter, and the peculiarly specialized version of the English language used to communicate effectively in the next.

A researcher or group of researchers will normally start to write a paper when research they have undertaken has revealed new information or findings that the authors believe (1) is reliable enough and (2) important enough to publish without further delay. The two concepts mentioned here, of reliability and importance, of course are critical to the rationale for any paper, and will be discussed in much greater detail later.

If research can be defined as the process of generating new knowledge, then a key step in sharing that knowledge with all those who may benefit from that knowledge in a credible way (unlike, say, tweeting about it!) is to validate its worth by achieving acceptance in a respectable scientific journal. If a piece of research is done that was worth doing, and yielded results that are worth sharing, then publi-cation in a peer-reviewed journal is the primary way of sharing this that scientists will consider seriously (perhaps previewed through glimpses of the work at confer-ences and other public fora).

Sometimes, the decision to publish may be taken for other reasons, such as a project coming to a natural end (or an unnatural one, when funding runs out) or researchers needing to meet the expectations of funding agencies. In addition, there can be personal pressures to publish which will drive the impetus to write papers. As will be discussed throughout this book, papers are the currency of cred-ibility for scientists and, when looking for new jobs or for promotion in current ones, or heading toward the examination of a PhD degree, publication lists will be scrutinized. This tends to focus the mind toward subjecting research to the rigors of peer-review with a goal of eventual publication of an impactful paper in a respectable journal.

Whatever the reason, at some stage attention and effort will switch from the lab or field to the business of serious writing. An initial consideration will be what to put in the paper, and it is critical to note that this will rarely mean "every single thing we did"; this needs to be guided by consideration of what the key messages to be conveyed by the paper are, and the corresponding parts of the work that convey that message. Diluting any paper by putting in every single piece of data just because they are available is never a good starting point, despite the hours that might have been generated in producing the data.[1]

Then, in terms of actually writing the paper, as mentioned in the previous chapter, it is relatively rare for a scientific paper today to be written by a single author, and papers today can have lists of authors from two (classically a PhD student and their supervisor or mentor) to dozens, hundreds, or even thousands (most commonly in biology or physics).

So How Do Papers Actually Get Written?

Before we discuss the actual process by which this takes place, let's consider a hypothetical scenario that illustrates what could perhaps reflect how this process would work. This is, frankly, unrealistic, but could represent what one might expect would be the process by which a paper is written, and presents a useful starting point for the exploration of the processes involved.

For this exercise, let's visit a hypothetical conference room, off a busy lab, anywhere in the scientific world. The Principal Investigator (PI) on a research project has decided, based on the results she has seen, and conversations with her team, that a paper is ripe for the writing. She has emailed everyone involved to summon them to this room and, at 8:30 in the morning on the appointed date, eight people assemble, including the PI and another senior researcher who collaborated directly in the work, two PhD students, each vibrating mildly with excitement with the thought of their first publication, a technician who played a key role in the analysis, the lab expert on statistical analysis, a collaborator from a nearby lab, and another collaborator who has flown in from overseas. They have all brought their own contributions, and notes, lab books, printouts from instruments, textbooks, key relevant scientific papers, and laptop computers soon cover the available surface of the table, and spill over onto the floor and nearby chairs.

A plan is agreed, with the PI acting as director, for the morning's activities. People crunch numbers, write up methods, draft an introduction or check and recheck statistical analyses before drawing up tables or figures, working singly or

---

[1] Note that this does not mean that data can be withheld or selected in a dishonest manner, which is a different consideration, to be discussed more in chapter 7.

in pairs. Multiple low conversations go on simultaneously as things are checked, ideas bounced around, and orders for coffee taken. A quick sandwich lunch is followed by a status report, as each team reports on its activities and all chunks are knitted together for the first time into a single file. Gaps and inconsistencies become evident and everyone goes back to their area for perhaps another hour before assembling another draft. Everyone then supplies suggestions for the arguments of the discussion, as papers are shuffled and references identified for citation, and this section is written in real-time, as people verbally edit text projected from the PI's laptop onto a white screen. The team works at a furious pace, with intellectual inputs flowing from all those present in a highly charged but deeply exciting atmosphere.

Then, the whole paper as reached at that point is printed and photocopied for everyone, and each takes an hour in silence with red pens to check and comment, while the PI drafts a proposed abstract and title, which are intensely discussed but agreed relatively quickly. The noise in the room then reaches a new peak as the entire text is scrolled through line by line and people fight over amendments and corrections. One of the team acts as the grammatical conscience, fixing commas and sentence structure, and moderating occasional arguments over the appropriateness of semicolons or dashes.

Then comes the sticky issue of the author list; the PI has thought this through carefully and makes a proposal, explaining her reasoning. The key resulting argument regards whether the technician involved should be a coauthor, and awkwardness ensues which is ended when the PI make the case that, during the past eight hours, they have made a significant number of intellectual contributions to the discussion and presentation of data, far beyond a simple role of performing certain analyses and being able to comment on these alone.

The paper is eventually declared done around 7 p.m., and the adrenaline rush brought on by the excitement of the work and the intellectual hurly-burly of the writing itself subsides, and yawns appear, not helped by the PI insisting that everyone wait and witness her actually submitting the paper through the selected journal's website, which takes twenty more minutes. Finally, worn out but deeply satisfied, the team retire to the nearest Irish bar for a well-earned drink.

This is a lovely story, and it is a pity that it is so unlikely to be true for the way in which the vast majority of scientific papers get written. Usually, in real situations, one author takes responsibility for writing a rough draft, which is circulated to the other authors, probably by email, and they add their sections or comments, until gradually a consensus emerges. Comments, in writing and verbal, add to the paper like layers on an oil painting, words get added and deleted, points get argued, and generally authors (often in different countries) reach a point where they are satisfied with (and/or sick of) it, and pass it fit for submission to the agreed journal.

It is rare enough to be noteworthy when scientists complete a paper in a single draft. It was said of the great Australian-born physicist Sir Lawrence Bragg (the youngest winner of the Nobel Prize ever, at 25, for his work, with his father, on using X-rays to determine the structure of crystals) that he was the only scientist who could take notes and data home with him in the evening and come in the next morning with a paper written out in longhand without blot or hesitation, publishable as it stood "rather like Mozart writing the Overture to the *Marriage of Figaro* in a single night."[2] More often, papers are polished in a long process by their writers before being submitted to a journal. The physicist Niels Bohr was a particularly obsessively careful writer of scientific papers. He would ask friends to read preliminary versions, and weighed their comments so thoughtfully that he would often begin over again; a frustrated collaborator once snarled to a colleague who had given Bohr a minor suggestion on a draft, prompting a seventh rewrite, that when the new version was produced, "if you don't tell him it's excellent, I'll wring your neck."[3]

Writing, it can be concluded, is not the easiest part of many scientists' work, nor their favorite, however much the end result may mean to them, their peers, and science. When a geneticist called Matthew Meselson provided sound experimental evidence that the theory of gene replication suggested by the work of Watson and Crick was correct, another colleague, Max Delbruck, and his wife carried him and his collaborator, Frank Stahl, off to a marine research station run by the California Institute of Technology and locked them into an upstairs room with two sleeping bags and a typewriter until they wrote the paper.[4]

So, in general, papers evolve through an iterative process of writing, criticizing, and rewriting, until a point arrives when all authors are adequately satisfied with its contents and presentation. It is of course debatable whether papers with enormous author lists are scrutinized by everyone present, and sometimes the same seems to happen even with smaller numbers of authors, which can lead to issues that we will revisit in the chapter on publication ethics.

What Does a Scientific Paper Say?

The modern scientific paper has evolved to communicate scientific information in what should be a highly efficient structure. There is relatively little difference in the structure of scientific papers in widely different fields of science, and certain key principles apply throughout publishing, which will be teased out in this and the following chapters.

In essence, every scientific paper must say a certain number of things, and provide good evidence to back up these things. These key elements of a scientific argument are as follows.

---

[2] Quoted in Judson, Horace. *The Eighth Day of Creation.* (London: Penguin Books, 1979), p105.

[3] Related in Crease, Robert P., and Mann, Charles C. *The Second Creation: Makers of the Revolution in Physics.* (London: Quartet Books, 1997), p21.

[4] This story is told in Judson, Horace. *The Eighth Day of Creation.* (London: Penguin Books, 1979), p192.

1. We, whose names appear atop this paper, have something to report.

As discussed already, a paper can have between one and many authors, and each should have a good reason for their name appearing, and be willing to stand over the contents. This is a key area relating to research integrity, which we will get to later in the book, but for now let's assume a list is proposed that is fair and inclusive. This is only the start of the problems, though, and the real jockeying for position can relate to the order in which the names will appear.

Conventionally, there are two good places to be, the best seats in the house. With casual disregard for biblical metaphors, the first shall be first, and the second shall be last; in between matters far less than either of these two prime spots. In most fields, it is accepted that the first-named author generally did most of the actual work, while the last-named is the one who directed the research. In a simple case, the first-named would be a PhD student and the last their supervisor. Last place should however be awarded on the basis of real direction and engagement in the research, not just seniority. For example, a department head or lab boss, unless genuinely deserving of authorship, does not deserve to have their ego massaged by being awarded this place; such practice is known as honorary authorship, and is regarded today as thoroughly inappropriate.

The value of being first author can be seen easily when it is considered how the paper will in future come to be referred to in the literature; a paper by Kelly and six authors will be known forever after as Kelly et al., with "et al." being derived from the Latin for "and all the rest."

Although principles as to the conditions that allow someone to be listed as an author of a scientific paper have been articulated and broadly internationally accepted (such as the so-called Vancouver protocols for authorship, which were first articulated by a council of editors of medical journals in 1978 and will be discussed more later), it is often the case that deciding who should be an author of a paper can become a fraught and ethically problematic discussion.

2. We will endeavor to present our findings and explain our work as clearly as possible.

Above all, a paper must be readable, and the reader must not have to struggle to understand what is being reported. The clarity of language must under no circumstance become a barrier, and is expected to be of a certain standard (although frequent readers of papers may agree that this standard is not uniformly met!). We shall discuss the language used in scientific papers more in chapter 4.

3. We believe that something to be an original and interesting contribution to scientific knowledge. In addition, we can support the above claims following our careful and thorough reading of the literature in the field, which we summarize (briefly) in the Introduction of the paper, where we recognize and cite the most important studies relevant to the paper.

Originality is key; a scientific paper is supposed to be a report of new knowledge. For a paper to be published in any scientific journal worth its salt, evidence must be provided that the knowledge is indeed new. To support this assertion, the authors must acknowledge the field of their research, demonstrating their awareness by citing the appropriate papers, and then outline precisely how their paper extends this knowledge and adds to it. In other words, the gap in our knowledge must be shown, and the way in which this paper plugs this gap must be demonstrated convincingly.

The classic put-down of a scientific paper says it all: "This papers contains things that are new and interesting; however, the things that are new are not interesting and the things that are interesting are not new."[5] Identifying whether the material in every paper is really original, or if the original material reported is in any way useful or interesting to the scientific community, is beyond what the editors of any journal in many cases can precisely judge. For this reason, evaluations of this sort are one of the key responsibilities of the referees enlisted to apply peer-review to submitted manuscripts; we will discuss the process of peer review in great detail in chapter 5.

Notably, originality is not in itself enough, even though it may seem a daunting task, particularly for a PhD student just starting their research, to produce knowledge which NO ONE ELSE HAS EVER REPORTED BEFORE!! However, originality, in many fields, is not that hard to achieve, and as a result there is a parallel requirement that originality be accompanied by a case being made for that hard-to-pin-down quality of interest or relevance.

To illustrate this point, consider my own experience, as editor of a dairy journal to which papers on all sorts of frequently obscure cheese types of other local dairy products are regularly submitted. A paper could describe the microbiology and chemical flavor profile of a particular cheese only available in a very remote geographical area, and this would (probably) be an original contribution to knowledge which no one has ever described before. However, in this case an argument could readily be made that novelty is not the same thing as originality, and no journal will publish research for which there is no clear case being made that there is a potentially interested audience, and which thus cannot be deemed to be likely to make any impact (which is probably one of the most prominent concerns in scientific research today).

Citation of papers, which refers to published work being subsequently referred to by later papers, is arguably the key currency of measurement in science today, whether applied to individual papers, summatively to journals (through their Impact Factor), or authors (through the h-index, a single number that combines measures of the number of papers they have published and the number of citations those accrued). To look at the issue of a paper with novelty but no broader significance strictly in terms of citations, a paper that no one cares about will not attract citations, and a journal that publishes such papers will suffer in terms of their Impact Factor, which will mean that authors are less likely to send their best papers, which hurts their Impact Factor, and so forth. This may seem like a

---

[5] A similar damning sentiment could be targeted at a paper that it "fills a much-needed gap."

cynical perspective, but it shows why being the first to find or document something is not enough to get published if the main long-term impact is a warm fuzzy feeling for the authors.

Originality, without self-indulgent novelty, is the key argument to be made by the authors. Originality can come in many different ways, whether new methods to new findings or new topics being studied, new conclusions being drawn or old ones being overthrown, or even old data being re-evaluated in new ways (for example, through meta-analysis of multiple previous studies to search for new patterns). Different papers may claim to be original in different ways, but their claim must be explicit and clear for reviewers to evaluate.

So, the Introduction to a paper must briefly review key papers in the field of the paper (without becoming a long review of the historical literature) such as to make a case for what is known already, what gaps there are in such knowledge, and why a paper that addresses this knowledge gap is important and timely.

4. We have used the most appropriate scientific methodologies in our work and, in addition, have provided sufficient detail on our methods and experiments so that anyone, who should so wish to do so, could replicate them precisely.

When scientists report something new and potentially of interest, the first question of their peers, the readers of their paper, is how they did the work, in terms of what methods they used. To take the findings seriously, expert readers (and, in the first instance, reviewers) must agree with the way the work was done, because they will not believe the paper's conclusions unless they are satisfied with the tools that have been used to underpin the argument.

Sufficient detail must be provided so that someone, if they so wished, could repeat the work in its entirety and thereby either build on it or disagree with its outcomes; this rule is straightforward and rigidly applied, as it represents one of the quality control points of research. Methodology in science is like the foundation of a building—if this is not solid and dependable, everything that is built on it is shaky. Conclusions and findings arise from data, and data are generated from methods and designs of experiments and studies. If the way in which the work was done is untrusted, untried, or simply inadequate (for example in terms of replication or safeguarding against the risk of uncontrolled factors influencing the outcome), then the data generated cannot be trusted and no subsequent conclusion can be relied on.

In some cases, there may be several different approaches to doing a certain piece of research, and researchers may be limited to what they have available, but in all cases they must justify their choice of methods such that reviewers and readers have confidence in the results they present.

Secrecy and coyness, which might be associated with many other aspects of human activity, are simply unacceptable. If the work is sensitive or secret, don't publish it, take out a patent, or do whatever you have to do, but if authors step up to submit a paper, these are the rules, and they must follow them. There have been cases where authors have sought to obscure the finer details of their work specifically to throw competitors off the scent,[6] and cases where it has been judged that to publish full details about exactly what they have done could actually be a hazard to mankind.

5. We have shown all experimental results that form the basis of the novel contribution we are making, in a clear manner so that the reader can decide if our claims are substantiated. In addition, we have ensured that our results are reproducible and consistent, and, where appropriate, used relevant statistical methods to verify this.

The heart of any scientific paper will be the Results section, where tables, figures, plots, images, or other artifacts are used to present the outputs of the research being presented. These are accompanied by a commentary in which the authors guide the reader through this evidence, drawing their attention to key features to be noted and how to draw conclusions from the processed data presented. In addition, in most fields, the results are buttressed and arguments prepared in advance by the application of appropriate statistical tools, to reassure the readers that the results are not a fluke and have a reasonable degree of certainty of being something to get excited about (while avoiding Mark Twain's assertion of their being three types of lies: lies, damn lies, and statistics). One of my favorite quotes in this regard is that a statistic is just a number looking for an argument, and this is essentially its function in a paper, to bolster and strengthen the authors' argument.

It pains me to have to point it out, but all the results that appear in the paper must be real. Not too long ago, it was assumed, perhaps innocently, that the likelihood of someone submitting a paper that contained anything other than genuine results was remote. In a biblical analogy, the scientific community lost its innocence in this regard beginning in the 1980s, and it is now necessary to point out the previously unmentionable. Authors of a paper must obey a new set of commandments, and the sins of data omission or commission, of fabrication, falsification, or plagiarism, must not be considered or enacted. The elephant in the corner has lifted its trunk and roared for attention, and the new era of ethics and integrity in scientific writing and communication more broadly will be explored in chapter 7. As a result of the soul-searching induced by such cases, the focus is more than ever on the assumption that all results presented are an honest representation of the real outcomes of real research, with severe consequences for researchers who abuse such trust.

---

[6] For example, a story of a chemist who blamed a typographical error, rather than competitiveness, for a reference to ytturbium rather than yttrium in a paper on his new wonderful superconducting material, in response to other researchers clamoring that the material did not have the properties claimed. This led to a wonderful 1987 headline in *Science* of "Yb or not Yb? That is the question."

6. We have attempted to explain the observations we have made, again by reference to appropriate prior knowledge, and have commented on the significance and importance of the findings for the field.

It is simply not enough to put results down on paper; it is expected that the authors of the paper contribute to the interpretation of the results presented, guiding the reader toward a reasonable understanding of the significance of the findings. Normal standards of rhetoric are supposed to be followed, and referees and editors must ensure that the authors are reasonable and realistic in their discussion.

Criticisms that are sometimes leveled at papers include practices such as HARKing ("Hypothesising after Results are Known"),[7] where researchers retrofit a hypothesis after the results of their study are known. There is also ongoing concern about studies being designed with insufficient statistical power (such as number of samples or replicates) to draw meaningful conclusions, and P-hacking (sometimes known as data trolling, dredging, or fishing), where data are collected or manipulated by applying multiple statistical tests until apparently statistically significant outcomes can be reported.[8]

Hyperbole in discussing one's results is certainly frowned upon, a tone of respectful modesty more in line with standard practice. One of the most famous lines of scientific literature, from Watson and Crick's legendary 1953 paper reporting the double-helical structure of DNA, stated, "It has not escaped our notice that the specific pairing we have postulated immediately suggests a possible copying mechanism for the genetic material." This has been described as coy in its understatement of the impact of their structure for the understanding of probably the most fundamental process of life itself, and stands as a splendid exemplar of the mark that authors must strive to hit.

7. The authors have all made significant intellectual, scientific, or practical contributions to the conduct of the research and/or the writing of this paper and take full responsibility for its contents. In addition, no individual has contributed to this work without due acknowledgment.

This is another area where scrutiny has come acutely to bear in recent years, in that journals are increasingly conscious of the ethical issues associated with authorship of a paper. As mentioned earlier, the reputation of a scientist depends in large part on their receiving credit for that to which they have made a genuine scientific contribution. This means that, where credit is due, this must be recognized by authorship, so that those deserving of such credit get the consequent benefit for their work and career, while others do not get unfair advantage by accruing undue credit through unearned authorship. As well as this, all authors have to accept responsibility for all or some of the paper should problems arise in the future,

---

[7] Kerr, N. R. "HARKing: hypothesising after the results are known." *Personality and Social Psychology Review*, August 1, 1998.

[8] It has been said that if you torture data sufficiently, it will confess to anything.

and so with credit comes accountability and responsibility. These factors can lead some of the most significant ethical issues a young researcher might encounter (for example, where a much more senior supervisor recommends authorship allocation that simply feels "wrong"), and this will be teased out further later in this chapter and in other chapters.

8. Finally, we will help the reader, or potential reader, by providing a useful title and abstract for our paper.

Apparently, you shouldn't judge a book by its cover, but a scientific paper is judged first and foremost by its title. Today, scientists in any field are confronted any time they log on to a journal website, consult a literature-searching engine, or even (unfortunately, increasingly infrequently) pick up a good old-fashioned paper issue of a journal, with dozens or hundreds of papers clamoring for their attention. It would take huge amounts of time to read every paper thus found to ascertain its importance or relevance to the researcher, so scientists learn to screen, and the first way they screen is on the title. The title is the authors' first chance to say, "Hey you, yes you, stop browsing and look at this, because this is good stuff and you need to read it."

I am sure marketing experts could teach scientists a thing or two about advertising their wares, because this is what the title of a paper must do. As with every aspect of the paper, outlandish or bizarre titles are not acceptable, and authors must work within far tighter constraints of what is acceptable than the world of advertising, where it seems that anything can, and usually does, go. More seriously, for a paper to make an impact, it must first be read and, for it to be read, it must attract the attention of those likely to appreciate it.

Another potentially useful analogy is that of marketing of movies. One level of advertising is based around the film's title and poster, which have limited space and opportunity to make an impact, yet must connect with the viewer and make them curious to learn more about the movie (i.e., playing together a role analogous to the title of the paper). The second level, assembled with more care and more demanding of the attention of the potential "customer" is the trailer (in this analogy, the abstract of the paper), where perhaps the filmmakers have two minutes (or 100–250 words) to make the real hard sell, and commit the unwary watcher/reader to paying to see the film/read the paper.

Paper titles must be not too short and not too long (perhaps ideally between 4 and 12 words, as a general rule of thumb). They must capture the spirit of the paper, and give the reader clear clues as to content, field, and approach (and yes, preferably, within 4–12 words!). Some journals will allow what are termed assertive sentence titles (ASTs), where the paper is boiled down to a punchline ("Compound X is a highly effective inhibitor of infection of monkeys by virus Y"). Others frown upon such dogmatic declarations that seem to broach no argument, and sometimes seem to flaunt normal standards of scientific caution, and prefer more neutral titles like "Effect of compound X on the virus Y in monkeys." Grammatical sophistication is not a major concern (once clarity is achieved), but subtitles may be permitted ("Virus Y in monkeys: compound X may be good news"), as may, perhaps surprisingly infrequently, be direct questions ("Does compound X inhibit infection of monkeys by virus Y?").

Whichever style is selected, usually after the journal's style guidelines or recent issues have been consulted to check the accepted norms, the authors pick their title; perhaps ironically, given its importance, a title is often selected last and often receives much less consideration than other, arguably less critical, parts of the paper.

Sometimes there is a direct correlation between the title and abstract, as when the abstract answers a question posed in the title. I can think of some recent examples that boiled this down to a spectacularly pithy abstract, as when a 2011 paper titled "Can Apparent Superluminal Neutrino Speeds Be Explained as a Quantum Weak Measurement?" had an abstract that read, in full, "Probably not."[9] However, this record for the smallest sized abstract was broken in 2018 by a paper that matched the title "Do Large (Magnitude ≥8) Global Earthquakes Occur on Preferred Days of the Calendar Year or Lunar Cycle?" with the one-word abstract "No."[10]

These eight points summarize the key objectives that every single scientific paper must achieve, in any field, and Table 3.1 summarizes how these are achieved in the different sections of a paper.

A frequently discussed element of the scientific paper is its intrinsically mis-leading appearance. Go into any library and pick up a journal and go through the first paper you see. What will be presented are an introduction; a list of the materials and techniques used; a set of results in the forms of tables, diagrams, or photographs; and a dry discussion of the meaning and significance of the results. There is no indication of the amount of work, the problems, the personal conflicts, the frustrations, the failures, or the feelings of discovery when the end results were achieved. None of these appear or can be gleaned from the clinical scientific style. In short, the process of science itself is missing from the paper. The chemist (and Nobel Prize–winner in 1981 for his work on chemical reactions) Roald Hoffmann said about this:

> Here is the journal report (scientific paper), a product of 200 years of ritual evolution, intended, supposedly, to present the facts dispassionately, with-out emotional involvement, without history, without motivation, just the facts. Well, underneath there's a human being screaming that I'm right and you're wrong. That endows the scientific article with an incredible amount of tension.[11]

---

[9] *Journal of Physics A: Mathematical and Theoretical* (2011) 44, 492001.

[10] *Seismological Research Letters* (2018) 89 (2A), 577–81.

[11] Quoted in Wolpert, Lewis, and Richards, Alison. *A Passion for Science* (New York: Oxford University Press, 1988), p24. Hoffmann was also a published poet and playwright.

TABLE 3.1 The Sections of a Typical Scientific Paper and Their Functions

Title	Short, informative, the hook, database-friendly
Abstract	The next hook, stands alone
Author list	Complete and ethical
Introduction	Set the scene; summarize the background; make the case; state the purpose
Materials and Methods	Don't believe me? Do it yourself
Results	All the key information, honestly presented in clear and useful figures and tables, appropriately analyzed, with commentary
Discussion	How it builds our knowledge
Conclusion	The take-home message
References	What is needed for reader and gives a sense of understanding of field
Acknowledgments	Funding; not-quite-authors

The Author List

One of the most potentially ethically fraught parts of any scientific paper is its author list. It has been pointed out several times already that inclusion on an author list allows one to include a paper in the publication list on their CV, which can be directly linked to career progression for any scientist in a wide range of ways.

It could reasonably be assumed that, to warrant authorship of a scientific paper, someone must have made a substantial contribution to the work concerned, but the big question here is what exactly constitutes "substantial."

There are two types of author who are of concern: honorary and ghost authors, the difference between which is easy to summarize. The former don't deserve to be authors, yet they are, while the latter deserve to be authors, yet they are not.

There are also examples of joke authorship, which I mention here just for purposes of, well, humor. I particularly admire the Nobel Prize–winning physicist called Andre Geim, who had a coauthor called H. A. M. S. ter Tisha, who was actually his pet hamster, while another paper (from 1978, on cell biology) includes as author Galadriel Mirkwood (named after a dog, in turn named after a Lord of the Rings character) and an American mathematician and physicist called Jack Hetherington apparently included his cat, F. D. C. Willard, on several influential papers in the 1970s.[12] There is also a famous example where the physicist George Gamow, writing a major paper on the Big Bang with a student Ralph Alpher, asked his old friend Hans Bethe to allow his name be added to the list of authors to allow it to read Alpher, Bethe, Gamow,[13] while another physics researcher took out his

---

[12] Tempted as I might be to recognize in this manner my dogs, Juno and Odie, for their dogged support, I have resisted to date, but will take the opportunity here to acknowledge their contributions to maintaining my spirits while writing.

[13] This is related, among other places, in Judson, Horace. *The Eighth Day of Creation.* (London: Penguin Books, 1979), p260. A similar impulse may have led to a paper titled "Observations of the 0-fs pulse" by Knox, Knox, Hoose and Zare, published on April 1, 1990.

frustration with having a paper rejected by publishing several papers authored by Stronzo Bestiale (Google it to translate!).

Turning back to more serious transgressions of authorship principles, a classical example of honorary authorship might concern a head of laboratory, department, or even institution who demanded that their name appear on every paper that came from that unit, irrespective of their involvement or lack thereof, simply because they are the boss. The most prolific author of all time is generally agreed to be a Russian scientist (Yuri Strukhov) who ran an institution for crystallographic analysis of compounds in Moscow to which scientists came from all over that country to perform analysis, on the agreement that his name appeared on ANY paper that resulted from such work. This was a good deal for Dr. Strukhov, leading to him publish 948 papers in the 1980s alone.

A ghost author, on the other hand, is someone whose contribution to a piece of work is genuinely sufficient to have earned them a place atop a paper, but who has for some reason been omitted, perhaps because they fell out with the other authors, left the group, or were simply not informed when the work to which they contributed was being submitted.[14]

An additional bizarre category of authors might be dubbed the "pay to play" authors, who are prepared to pay significant sums of money to have their name appear on a paper in a prestigious journal, knowing that the value to their careers of such an appearance is worth the price they will pay.[15] This was first spotted when some papers accepted into top journals had, after acceptance but before final publication, several new authors added, who it later transpired had been offered this for cash; this is one reason why most journals will now not accept changes in authorship after first submission of a paper. In 2019, it was reported that a website in Russia had brokered authorship for more than 10,000 authors for prices starting at around $500; they are apparently continuing to offer this service despite much legal pressure to cease and desist.[16]

The closest to a set of internationally agreed principles on authorship of scientific papers were agreed by a set of editors of medical journals (ICMJE, International Committee of Medical Journal Editors) in Vancouver, Canada, in 1985. The so-called Vancouver protocols state that authorship should be based on the following four criteria:

- Substantial contributions to the conception or design of the work; or the acquisition, analysis, or interpretation of data for the work; AND
- Drafting the work or revising it critically for important intellectual content; AND

---

[14] Apparently, after Strukhov died, his name continued to appear on papers for some years, a rare example of an honorary author becoming a ghost one.

[15] One example of this was described in "China's publication bazaar" *Science*, 342, 1035–1039 (November 29, 2013).

[16] https://retractionwatch.com/2019/07/18/exclusive-russian-site-says-it-has-brokered-authorships-for-more-than-10000-researchers/

- Final approval of the version to be published; AND
- Agreement to be accountable for all aspects of the work in ensuring that questions related to the accuracy or integrity of any part of the work are appropriately investigated and resolved.

The detailed Protocols (which can be found easily online) also helpfully indicate the type of contributions that warrant acknowledgment rather than authorship, including purely technical contributions, acquisition of funding, or supervision of the research group alone.

When faced with tricky discussions on the right way to resolve authorship disputes, these principles represent a good set of rules on which to fall back. Journals adhering to the Protocol will ensure its application by, for example, insisting that submissions be accompanied by letters signed by all coauthors, so that someone's name cannot be included on a paper without their consent or agreement that they meet the criteria for being so included. In journals not following the protocol strictly, one author can submit on behalf of all coauthors, leading to potential problems.

Of course, agreeing who deserves to be an author leads to a secondary and also potentially thorny issue, which is the order in which those names should appear. As mentioned earlier, it is generally accepted, in most fields, that the first author is the person who has done most of the actual practical work, and this pride of place is reflected by the fact that, in common subsequent forms of cross-reference, they will be the only name visible, as the paper is referred to as "Kelly et al. (2018)." Some journals will allow, where the contributions of two authors are of such similar magnitude as to make demoting one into the haze of those two short Latin words unfair, joint first authorship, such as "Kelly and Murphy et al. (2018)."

The next best place to be an author list, after first, is generally last, as the last author is generally the person who has supervised or played the most senior role in the work, and so last place is a recognition of this significant contribution.[17] This may be easy to identify, but in some cases students, for example, might have cosupervisors. In such cases, while it may be easy to identify which of those has had the obviously larger input, this may not always be as evident, and so open discussion of these matters is critical, as arguments can often arise where a decision on a proposed order of authorship is reached without all parties having had their input.

After first (or joint first) and last, the order of authors in between is probably not of huge concern in most cases, but generally would decrease in overall level of input from the start of the list on in, rising again sharply at the end.

Where there is only one author (perhaps most commonly encountered in mathematics) such issues will not arise, while the next simplest situation is a graduate student (first) with one supervisor (second, and last). However, in most cases,

---

[17] In some fields, such as economics, author lists may be in alphabetical order, but that is not the case in most scientific fields.

there will be more than this, and as mentioned earlier in some fields such as genomics or particle physics there have been papers published with hundreds of authors from multiple institutions and countries ("hyperauthorship"). I do not imagine huge arguments arise about any place there other than first and last in these papers, but do wonder how many authors directly contributed to the writing of those papers, and how that process was managed (an example will be considered in the next chapter).

The Heart of Any Paper: The Centrality of Data

Francis Crick and James Watson articulated in the 1950s slightly different visions of what was called the Central Dogma of Molecular Biology, which stated, in summarizing the flow of genetic information, that deoxyribonucleic acid (DNA) makes ribonucleic acid (RNA) makes protein. The details of the model are not the focus here, but the idea of a flow with a particular direction has an analogy, for me, in scientific research, which is that:

Methods Make Data Make Papers

A paper will only result from the availability of a set of good data, and good data will only exist where the methods used to generate them are valid and reliable.

The core of any piece of scientific research is the methods used, as if these cannot be relied on or are open to doubt, then their outputs are flawed from the outset, which is why the Materials and Methods section of any paper must be completely open about what was done, and any experiment or study should not be undertaken without a critical determination by the researchers concerned that the work will be done in the way that is most likely to be determined later to have been the best way to have approached the research in question.

Then, if the methods are in place and reliable, they are used, in one way or another, to generate data, and if these are interesting and lead to new information they form the basis of a paper (or thesis), being molded from multiple forms of raw data (printouts, images, handwritten data) into a set of figures and tables that demonstrate the key outputs and findings of the work, and represent the basis on which the authors' conclusions have been reached.

Without data, no output can be communicated, and in between the selection of the methodologies that generate the data and the construction of figures and tables lies a key skill of researchers, which is the management, archiving, and protection of data. A good researcher should become extremely professional and even paranoid about the organization and protection of their data. In my own university alone, I have seen every possible crisis that could be envisaged damage or destroy researchers' data, from fires to floods to computer hard drive failures, physical loss of USB keys (which pack more and more information into smaller and smaller objects), and viral infections of hard drives. Good researchers will have ingrained instincts about avoidance of such emergencies, such as routine backup procedures (physical or cloud-based) to keep multiple copies of critical files (while avoiding any confusion as versions begin to proliferate).

Of course, today, many institutions, funding agencies, and even journals have strict procedures and demands about the responsible open-access archiving of data, to ensure that they are available for fixed periods of time, for reasons including their availability in case of queries arising about the work in the future. Raw data may also be expected to be made available online for some studies, to allow researchers greater access to fuller information than that available in the paper itself.

An example of this was operated by the European Commission as the Open Research Data (ORD) pilot scheme, through which participating projects developed a Data Management Plan where they specify which data they will make available openly, and since 2017 that has been extended to the default position for all work funded under its huge Horizon 2020 research funding program.

The Use of Figures and Tables

Figures and tables are typically the core evidence displayed in a paper, having evolved as an effective way to summarize and present scientific data in a manner in which they can be evaluated and conclusions drawn. It could be joked that the perfect results section of a paper reads simply "The results are shown in Table 1," but in reality each table and figure must be accompanied by text wherein the authors assist the reader in interpreting and extracting key messages from what is shown. The narrative must guide the eye to the key points and features of data, without repeating in words what has been shown in the table or figure. The text should also avoid sentences that are simply signposts to a figure or table and still require a separate description—sentences like the example just given are the textual equivalent of empty calories, as they have no information content. References to tables or figures should be integrated into sentences that point out what the reader is supposed to see there, such as "There was a strong correlation between dosage of Y and level of X after treatment (Fig. 1)."

One key question to ask of any figure or table is whether it actually needs to be there or not. As a rule, if it can be summarized in words in a couple of sentences then the presence of a space-filling table or figure is not warranted. I have sometimes seen bar charts with two or three data points the information content of which could be stated in a single sentence!

If a figure or table is deemed worthy of inclusion, then, the next principle is not to treat this as a challenge to artistic or design skills and look for the most intricate way in which this could be prepared. Rather, each should be as pared-back and

minimalistic (in terms of flourishes like unnecessary lines, grids, or other orna-
mentation) as possible, to avoid distracting from where the focus should be, which
is the data presented.

Sometimes, there can be a choice as to which format is best for a set of data,
in terms of a table or a figure. A key primary consideration here is that the eye
and brain will find more impact from a graphical presentation than a tabular one,
and someone reading a table and comparing values is essentially trying to build a
mental picture of relationships and trends—giving a figure just saves a step here.
As we will see in chapter 8, this is even more critical when time is of the essence in
a presentation where a message has to be conveyed with impact in a limited period
of seconds.

Figures are good at showing relationships and trends, and getting these across
in a rapid and economical manner. However, if the key is for readers to see and
have access to absolute values, or if there needs to be significant inclusion of statistical analysis and evidence, a table will be a better choice.

In designing figures, the issue of use of color must be carefully considered.
Sometimes, a figure such as a photograph or schematic simply wouldn't make sense
unless it was in color, but it must be borne in mind that if a journal (still) prints
issues in hard copy, they will levy a charge for the printing of color images, even
if they do not otherwise apply page charges. As I said, for some images this will
be accepted as inevitable, if that is the policy of the journal chosen as best for the
material, but designing a plot and deciding to make lines and symbols different
colors simply because it looks nice is not sufficient justification for the payment that
will be required. It should also be borne in mind that some readers may like to make
printouts or hard copies of papers, and may not have color facilities for doing so.
So, use color where needed, and not just for aesthetic reasons. In some cases, in my
own journal, authors elect to have color copies of their figures in the online versions
of the paper, but black and white in hard copies, and this can work too.[18]

Overall, figures should be as free of unnecessary frills or complexity as possible,
including grid lines or other design features if these are not absolutely necessary,
and also distracting features like 3D effects.

Each figure or table must be accompanied by a descriptive and clear legend
or label, which is numbered sequentially (to allow cross-referencing in the text)
and explains what is shown, but does not discuss this (that will be in the accompanying text). It is usually said that such legends should be stand-alone and self-
explanatory (in terms of being able to read these without too much cross-checking
of text), but this can be hard to achieve; at the very least, all abbreviations should
be explained so that these do not need to be checked independently to understand
what is shown. A well-designed table can also achieve a lot through the use of

---

[18] Another example of a publishing practice that will presumably soon be of historical interest only
when the idea of printed copies of journal issues becomes (sadly) obsolete across most or all journals.

footnotes, to explain abbreviations or codes used, indicate missing or omitted values, or describe the statistical analysis applied.

A final point is that authors must be sure they actually have the right to include any figure or table if that has appeared previously elsewhere. Even if the authors created that content previously, if the journal or other publication in which it appeared is not open-access, they need to seek permission to use or reuse it in the new publication, and indicate that in the appropriate legend.

The Role of Acknowledgments

The Acknowledgments section of a paper has a number of functions, the first of which is to thank and note individuals whose contribution to the work is worthy of noting, but not sufficient to meet the criteria for authorship. This might include those who made specific suggestions, whose contributions were strictly technical (rather than intellectual) or who undertook critical reading of manuscripts, for linguistic or scientific reasons.

A critical part of the acknowledgments is the declaration of financial support for the work described, and funding agencies will typically have standard wording that they insist on being included in papers arising from work they have supported. Reasons for this include the ability to justify the ongoing award of funding from wherever they receive the funds they disburse, on the basis of showing the productive outcomes of previous such allocations. Such declarations of sources of funding also make transparent any possible implications of conflicts of interest, as they allow readers to judge whether there are any factors that may have affected the independence of the researchers, and hence biased their work in any way.

For a hypothetical example, a paper showing the positive benefits of a drug made by a particular company might be read with a different perspective from readers if they knew the work was sponsored by that company. For a far less hypothetical example, one of the serious issues relating to the infamous *Lancet* paper by Wakefield et al. in 1998 on the possible relationship between administration of the MMR (measles, mumps, and rubella) vaccine and autism on childhood was the fact that the paper did not declare that the work had been funded by a group of lawyers seeking to build a case against vaccine manufacturers on this very basis.

Some papers have had acknowledgments that appear on the surface to be innocuous but yet turned out to be highly controversial, including the famous paper by Watson and Crick on the structure of DNA. Their acknowledgments, which actually represented a significant portion of the word count of perhaps the shortest paper to ever spawn a whole field (molecular biology) are worth reproducing in full:

We are much indebted to Dr. Jerry Donohue for constant advice and criticism, especially on interatomic distances. We have also been stimulated by

a knowledge of the general nature of the unpublished experimental results
and ideas of Dr. M. H. F. Wilkins, Dr. R. E. Franklin and their co-workers
at King's College, London. One of us (J. D. W.) has been aided by a fellow-
ship from the National Foundation for Infantile Paralysis.

The reader might instinctively assume that those acknowledged appear in order
of descending order of magnitude of their contribution (with that of Dr. Jerry
Donohue being the most significant), which was later acknowledged (particularly
after Watson told the story of the discovery quite openly in his book *The Double
Helix*) not to be the case. In addition, the wording around the contribution of
Rosalind Franklin could reasonably be interpreted as, being generous, a very convoluted arm's-length understatement of the circumstances under which Watson
and Crick came into the possession of key data (in the form of an X-ray crystallog-
raphy photograph) which led to the construction of their double-helix model, and
which were provided by Franklin's erstwhile colleague Maurice Wilkins without
her knowledge.

Acknowledgments can also be a place for some humor or personality to shine
through, should the editor and publishers permit same, and there have been examples of marriage proposals appearing here, or authors thanking a huge range of
supporters or colleagues, or simply saying hello to Jason Isaacs.

Giving the Reader More: The Potential of Supplementary Material

With the advent of electronic availability of scientific papers, the opportunity was
quickly acknowledged to make additional (supplementary) material available for
any published paper than would have been the case for paper-based publication.
Today, it is common to find on a database a paper for which the authors have
uploaded additional figures, text (such as more details on Materials and Methods
that is useful but not mission-critical for the paper), images, video clips, background information or data sets, which are not critical for the understanding or
reading of the main paper but allow those readers who wish to delve deeper access
to a fuller set of information. This is potentially very useful for both readers and
authors, and broadens and deepens their engagement around the paper in question.

Having now discussed the structure of the scientific paper, we could envisage it
as a template or series of headings, for which we know the function and objectives.
Of course, these now need to be filled in with words, and so we will turn next to the
use of language in scientific writing.

{ 4 }

The Scientific Voice

Scientific writing is different from any other form of English writing, if for simplicity we consider here only those journals and articles published in that language. It has little in common with, for example, the forms and rhythms of poetry, fiction (high- or lowbrow), legal writing, preparing technical manuals, or even, arguably, common speech or writing.

It perhaps shares most in common with journalism, in its ideally cold and rational presentation of facts in a manner hopefully intended to educate the reader, with an added element of editorial comment. Personally, I have no greater heroes than journalists of courage and commitment, particularly those who put themselves in harm's way (which may today be closer to home than formerly) to bring the facts as they see them to us, comfortably sitting in our armchairs, little appreciating that for every tank firing or gunman scuttling for scant cover that we see on television, there is an unarmed journalist risking their life so that these images may cause us even slight pause in our channel-hopping. It may seem trite or insulting to such heroes to compare them to scientists working, in the majority of cases, in safe warm labs around which bullets ricochet only seldom, but I argue that the analogy holds.

Scientists and journalists work at the edge of what we take for granted, and strive to teach us about that which they have just found out and which they believe we need to know. There is even an analogy between scientific writing and high-quality political journalism, so important today, in that both seek to observe a situation or phenomenon, and then critically analyze it and present its meaning and interpretations to the world in as impartial a manner as possible.

Returning to scientific writing, whatever benchmark we use, it is inarguable that a certain style prevails today. To learn to write impactful scientific papers, a scientist must first learn to write in this style and voice. It may seem bizarre to suggest that the way in which a paper is written is as important as what it has to report, but that is what I believe. The reason for this is simple; the writer may have discovered something of huge importance to their field, or mankind in general, and worthy beyond question of a Nobel Prize, but for this discovery to be rewarded in an appropriate manner, a few things must happen:

- The finding must be presented in a location where those who will either recognize its significance or benefit from its conclusions will see it (i.e., there is no sense in publishing the cure for something important in the *West of Ireland Gazette of Mildly Interesting Trivia*, circulation 27, including 2 sheep);
- The finding must be presented in such a manner that those who read it must be able to appreciate immediately its significance, and those of expert leanings must be able to judge its scientific quality.

The first point is principally about judgment, as was discussed in more detail in chapter 2; the second revolves around the fact that a hugely important characteristic of scientific writing is *clarity*. This is where scientific writing diverges from most other forms of English discourse, as floweriness, imagery, alliteration, onomatopoeia, elegant variation, and myriad other tools of writers with literary ambition must be consigned to the dustbin.

The risk of misinterpretation or misunderstanding should be essentially zero. A sentence such as the following could never appear in a scientific paper:

"I saw a man using a telescope."

In this case, the writer of the sentence could have been using the telescope, or the person they saw could have been using it. Either interpretation is possible, and the reader is left either unsure or guessing, neither of which is an acceptable response to a scientific sentence.

Mastering the full toolkit of writing tools needed for the standard of scientific writing that is expected is not necessarily easy for young scientists to achieve, and writing in this manner certainly does not come easily to all. The skills and tools, mental and physical, that are required for good writing are not necessarily the same as those required for good laboratory skills, manipulation of instruments, or creative design of experiments.

It also does not always follow that students achieving high grades in scientific subjects at school or university will equally excel in English language; in some cases, the opposite may even hold true, where, at some earlier branch point, an individual's perceived strengths or interests led to a focus on science, which reduced the exposure to writing criticism, literature, and essays more traditionally associated with the humanities. Hence, writing must be learned and practiced, and feedback from more experienced practitioners of the art of scientific writing is one key factor for the development of the requisite skills.

The Value of Criticism and Self-Criticism

A key aspect of scientific practice is criticism; findings, proposals, suggestions, ideas, and writings are automatically questioned by one's peers and superiors, to evaluate their validity, identify their weaknesses, and improve their ultimate value.

Science can be a tough life at first for the weak-hearted, as their great ideas and hard-wrought writing are torn asunder by their supervisors, colleagues, referees, editors, and even complete strangers at conferences. The rule is simple; if you think you have the right to say something, then everyone else has the right to attack that and, if it survives this red-hot glare, it is stronger and more credible, while, if it does not survive, time has been saved and the work moves on.

Francis Crick, codiscoverer of the DNA double helix, once said of criticism that the first rule of criticism is to always find something good to say, before you point out the problems (for example, "I liked your choice of font and margins"). Scientific criticism should be aimed at the message, not the messenger, but that distinction can be hard to detect until accumulated scars build a tough skin. Writing is an obvious target of this criticism, as facts are presented in a form that can be analyzed at length, usually in the absence of the author, lending distance and impersonality, and perhaps emboldening greater depth of evaluation. Young scientists should accept this as part of their learning curve, and get practice early and work hard to improve their writing. For postgraduate students, feedback should be sought very early, and not left to the feared demolition of their entire thesis by a red pen, with the resulting document looking like it has been through a blood-bath.

By the time a thesis comes to be written, a student should have mastered the tools of the trade sufficiently to produce a well-argued document where the reader almost does not notice the writing and can get straight to the heart of the science.

The keys to becoming a good scientific writer, in my view, are thus practice, development of a critical eye, and the acceptance of regular constructive feedback. One must write as much as possible, become aware of good and bad writing (and how to achieve the former and avoid the latter) while reading, and look for every opportunity to gauge the response of readers to the efficacy of your writing. Without a good feedback loop, mistakes are repeated and become ingrained, rather than iteratively being corrected and writing gradually improving as rough edges get progressively smoothed down.

If I look at my final-year undergraduate dissertation, and my PhD thesis 5 years later, and then some recent writing pieces (this one included!), I can track an evolutionary pathway. This has hopefully converged on a standard and style that I will never regard as perfect or without room for further improvement, as ongoing review reports regularly point out,[1] but I am satisfied as being better for sure. The feedback loops that led me to this place are enormous and convoluted, representing the combined inputs of hundreds of others, from my first supervisors to my coauthors, collaborators, and students, as well as countless anonymous referees, plus the learnings from reviewing and editing hundreds of other papers. Each of these inputs represented a nudge toward or away from some facet of

---

[1] The week of writing this, I received a report on a review article (a type of article the usefulness of which is directly proportional, in my view, to the quality and clarity of its writing, even moreso than a conventional research paper), which I had spent a lot of time on, that offered the moderately satisfying verdict, "This study is good written."

writing, brought something to be reinforced, emulated, or avoided, and collectively have led me to the approach I use today. I am a great believer in the idea that you should be able, at the end of every day, to identify something new you learned that day, and many of these daily learnings have revolved around writing.[2]

In terms of feedback and criticism, the first and most harsh critic should be the writer themselves. Why would someone give a piece to others to labor through and correct when they have not taken the time to do the first pass themselves? Some writers will, on completion of a sentence, go back to the start of that sentence and polish it until shining before moving on, while others might tend to wait until large blocks have been complete before rereading, but reread and edit the writer must, whether on-screen or, in my view more effectively, on paper with pen in hand.

I will often edit and re-edit documents to a level that can border on the obsessive,[3] and eventually get to a stage that feels like the final minute of a haircut, where the barber or hairdresser makes a few final snips or tweaks. At this stage I am apt to scoot through the text, changing or deleting a word here and there, and know that if left for long enough I could get caught in an immobile loop (the length of which is proportional to how nervous I am about the piece in question), until eventually I convince myself to abandon it to the reader/reviewer/editor's mercies. A colleague of mine in University College Cork[4] once said that there are two types of thesis, perfect and submitted, and the same can apply to any piece of scientific writing—forget about aiming for the former and get to a place where you feel the latter is comfortable.

Overall, I find the process of rereading and editing to be far less fun than writing new text, and far less motivating, but it is critical to do so in order to hand the best version of the material to the next person to read it. In the case of my first book, that was certainly true, and a peculiar experience after months of on-screen editing and correcting was to work on a printed hard copy draft. That made it seem like a totally different manuscript and exercise, far less personal and representing more of a break compared to writing and editing in the same medium (keyboard and screen); this created a helpful distance, which is vital for good critical editing.

Aiming for Invisibility

Good scientific writing should almost be invisible, in that the reader should be carried along by the flow of the piece, focusing on the content as guided by the

---

[2] What then has been my greatest trend over 20 years of practice and review? It's simple—my sentences have gotten shorter. If plotted on a curve, I am positive that average sentence length will have decreased, if probably not linearly, with time. If readers might find this hard to believe, you should see my earlier papers!

[3] Probably a minimum of two full edits on any paper of which I am coauthor, and frequently far more.

[4] Thanks to Dr. Tom Carroll of our School of Mathematical Sciences.

invisible hand of the writer. A clumsy grammatical construct, an ill-advised typographical error or an ill-placed comma can be a harsh interruption to the experienced reader/writer, like fingernails scraping down a blackboard, jarring them away from the content and undermining their confidence in the writer and consequently what they have to say.

George Orwell, in a 1954 essay called "Why I Write," put this beautifully when he said that "good prose should be transparent, like a windowpane." Similarly, good scientific writing should not obscure that of which it presents a view, and should not be a barrier in itself to understanding and engaging with the subject.

An analogy can be found in movies, such as Peter Jackson's astonishing *Lord of the Rings* trilogy, where absolute immersion in the fantasy worlds depicted in the films depends entirely on the special effects, sets, and costumes being so believable that one can almost believe they are real; a false note or poorly done effect will wrench the viewer away from the story.[5]

So how does one achieve this grace, subtlety, and invisibility in writing? That will be the focus of the rest of this chapter.

Some Basics of Good Writing

The first thing that must be accepted is that scientific writing requires a good grasp of grammar. This is unavoidable, in my opinion, although this may cause dismay among those young scientists who gravitated to scientific subjects precisely as something apart from such foul things, with harsh memories of school lessons where rules must be learned by heart, essays written, and boring poems half-heartedly analyzed.

Grammar is the key to flow, and flow is the key to good scientific writing. Opinions as to the best style of writing vary, but I am firmly of the opinion that scientific writing should flow as speech does and that, if a piece of writing sounds good when read aloud, it will read well. This, I believe, is the key to good writing. It then follows that, just as good speech is highly dependent on such things as emphasis, rhythm, pace, and logical construction of sentences, so too is scientific writing.

Flow is all about rhythm, and grammar provides the metronome that tells the reader when to pause or flow, to separate ideas and join points, and thereby underpins the rhetorical construction of what is to be communicated.

However, I believe that scientific writing does not require young scientists to devour books of English grammar,[6] and relearn rules gratefully abandoned or

---

[5] Such as a joking reference to golf did for me in the later Hobbit trilogy!

[6] I would rather suggest here some perhaps unorthodox choices, such as Lynn Truss's *Eats, Shoots and Leaves* or Stephen King's *On Writing*. Neither is an uncontroversial textbook on writing, and neither relates specifically to scientific writing, but both will deliver a broad awareness of the importance of words, grammar, punctuation, and syntax in an enjoyable manner.

forgotten many years before. Rather, grammar can be learned from remembering a few simple guidelines, and then by practice, practice, and more practice.

I was going to call them rules, but that sounds too prescriptive. There is no absolute right or wrong in scientific writing, much as I or others may like to suggest there is. There can often be several equally acceptable ways of saying something, but also many (perhaps more) ways of expressing it badly. I claim no expertise greater than that which could help to avoid the latter, while perhaps hopefully ending up at the higher end of the former. This has been learned through perhaps a quarter of my working life (as a very crude estimate) being spent with a red pen in hand, editing and reflecting on my own writing or that of my students or the authors of papers I am refereeing or editing. This so-called expert eye is then kept on its toes (with apologies for mixing my anatomical metaphors) by ongoing training on a wider set of nonwork samples. I (in common, I suspect, with many for whom communication of one sort or another forms a huge part of their work) constantly and even unwittingly edit and critique everything I see, in newspapers, online, or spewing from the mouths of politicians or other public speakers.

Readers who are experienced writers will almost certainly disagree with some aspects of my advice, which is fine (as they say, all generalizations are bad), but basic principles discussed here are I believe mostly unarguable (while bracing for the inevitable arguments that will ensue). At the end of the day, clarity and readability, and resulting invisibility, are surely uncontroversial goals that a scientific writer must aim for.

PRINCIPLE 1: LISTEN TO WHAT YOU WRITE

It is vital to develop an ear for your own writing, as you must learn to judge the clarity of your own writing first and be your own harshest critic. Many an inexperienced student writer has delivered poorly written material to me, and all I have to say to demonstrate to them what is wrong is to tell them to read it aloud, putting pauses where commas are and obeying a few other basic principles; usually, the reader only gets halfway through a mangled construction before seeing the light. Principle 1 also means that you should never let someone else embarrass you in this way, so you should do it to yourself first. If it sounds right, it probably reads right.

PRINCIPLE 2: SYNTAX IS NOT A RELIGIOUS LEVY

Syntax is the construction of well-ordered sentences through the effective deployment of word and phrase order. A good scientific sentence should yield its meaning in a single reading. If it needs to be reread to understand it, then it is a bad sentence. If, after multiple readings, the meaning is still unclear and the reader must take an intuitive leap and hope for the best, then it is a very bad sentence.

In the following example, from the kind of writing I would regularly edit, it is not clear whether the whey samples or the milk samples were diluted and centrifuged, leading to unacceptable ambiguity.

Murphy, Kelly and Smith (1998) also proposed an electrophoretic method for the determination of cations in milk, including whey samples obtained from skimmed milk samples which were diluted and centrifuged.

In the following sentence, word order is the problem:

Although the determination of cations in milk samples is widely described, using techniques like spectroscopy, very few applications of CZE to the analysis of metal ions in milk were published.

The placement of the fragment "using . . . spectroscopy" is confusing here, as it follows and so implicitly relates to the describing, rather than determination, and so should be moved to immediately after "samples" (or, even better, after "milk" with "samples" being deleted, as this is probably superfluous in this sentence).

Finally, this sentence illustrates another key thing to watch for in scientific writing, which is appropriate use of tense. In general, facts are presented in the present tense ("DNA is a double helix"!), while work that has been conducted by the authors or others is presented in the past tense. In the example above, the verbs of concern are "is," toward the start, and "were" right at the end, which should be replaced with "has been" and "have been," respectively.

PRINCIPLE 3: BE A SLAVE TO THE RHYTHM

A key part of good writing, like music, is rhythm. The comma is your best friend in this regard, backed up at increasing length-scales by the full stop, and then the paragraph break. The comma is the ticking metronome of grammar with, in my opinion, the ability to make a sentence beautiful, or as ugly as a warthog with extra warts. It is frustrating to see a misused comma,[7] and appallingly common in modern media, but that is no excuse. Just because it is good enough for your local newspaper or some random website or tweet does not make it good enough for the scientific literature.

A comma's roles are simple; it tells you when to pause, and it separates points or arguments. It also helps us to avoid the kind of confusion seen when a chef was once mysteriously reported in a headline to "find inspiration in cooking her family and her dog."

When in primary (elementary) school in Ireland, one of the first rules of English I remember learning is that, when you come to a comma, you take a breath. This is one of the few things I learned from a nun that I still use every day. If you read

---

[7] While acknowledging that such a statement invites all my dear readers to point out all the unbeautiful ones in the text you have before you, but I have done my best.

a piece aloud or whisper it into your mind's ear and come to a place where you pause, then there shall be a comma. Conversely, if it feels awkward to make yourself pause where a comma has been placed, then get rid of it.

Most commonly, we encounter the comma separating items in a list where these items are described by single words or clusters thereof.

> "The principal elements identified in the sample were carbon, hydrogen, sulphur and oxygen."

In some cases, the "and" may be preceded with another comma, as in "sulphur, and oxygen." This is known as an Oxford, Harvard, or serial comma,[8] and may be used either because of a stylistic preference, or because to do so makes the sentence less ambiguous in its meaning, being a very firm means of separating the list items and making sure the items on the list could not be read as being more explicitly related than they actually are. This avoids such problems as the news headline that once read "Top stories: world leaders at Mandela tribute, Obama-Castro handshake and same-sex marriage date set" or the possible apocryphal book dedication to "my parents, Ayn Rand and God." In any case where the lack of an extra comma could potentially create confusion (or scandal), the Oxford comma is a grammatical safeguard, but where the list is unlikely to entangle the unwary reader it is probably unnecessary.

We also encounter the comma acting like a lower-key version of a bracket, surrounding a piece of information that can be removed and leave that which remains to still read clearly. This is a key way to achieve brevity and conciseness in writing.

> The rings of Saturn can be seen clearly with a small telescope. They are made from tiny particles of dust and rock.
> The rings of Saturn, which are made from tiny particles of dust and rock, can be seen clearly with a small telescope.

There are only two words in the difference, but the road to verbose hell is littered with single words and pairs thereof.

We also use commas to build in a pause within a sentence that combines two separate points but that we want to run closely together. In such cases, however, we need to add a linking word, typically "but," "while," or something similar.

> The sample reached a higher temperature during the reaction than expected, but this was due to the exothermic nature of the phase transition that occurred.

---

[8] The Oxford comma was the unexpected subject of a great song by the band Vampire Weekend from 2008, possibly the only song I can think of so devotedly dedicated to an element of grammar (many songs about serial killers, not so many about serial commas). I also was persuaded by my editors to include one in the title of my first book, and could not question the style of a house whose name included the city in question!

There are also some words that, even if found right at the start of a sentence, need to be followed by a comma to build in the dramatic pause that is required before the big reveal to come. Words like "however" or "thus" are the linguistic equivalent of tapping a wine-glass and clearing a throat—they show that the writer means business and has something big to unveil once everyone has focused their attention on the topic in hand. These words also serve to link sentences effectively while showing a clear flow of logic from the previous sentence or point, either to confirm or explain ("thus") or to draw a contrast or change direction ("however"), and again work better paired with a comma.

So, commas build rhythm into writing as they do in speech, telling the reader when to slow down, even briefly. In editing scientific writing, I probably add, delete, or move commas more than any other piece of punctuation.[9]

Full stops then create more definite breaks and pauses, between sentences rather than parts of sentences, and at the next level of breaking up text and argument we meet the paragraph break.

I believe that paragraphs simply make text easier to read and less intimidating. If I am editing a document and encounter a full page of text without a paragraph break, I groan (sometimes aloud) as I know a trudge is ahead across rocky terrain without pauses for rest, and my red pen will start to twitch as I feel an irresistible urge to carve footholds into the sheer rock face. It is simply easier to read text where there are these little mental break-points built in where the reader knows the text is going to change angle or topic even slightly. I then look for these even subtle swerves in the road and stick in a mark to add a paragraph break.

This might be a rule occasionally broken in fiction or more stylistic writing, as when the singer Morrissey began his autobiography with a single four-page-long paragraph, but this kind of thing would really not work in scientific writing, where we would be asking when we can take a mental break, and how soon is now?

One simple principle of paragraph construction is that they should be shorter than a page but, at the other end of the scale, (usually) longer than a single sentence. A second is that each should have an invisible title that describes what the paragraph is about, which is different from that of the paragraph that came before it, and that which follows. We don't put in these titles, but should be able to imagine that our paragraph is coherent or self-contained enough that it could be labeled thusly.[10]

In formal terms, a paragraph should begin with a topic sentence, which gives a sense of what the paragraph is about. The sentences that follow within that paragraph then expand, explain, or describe further the point(s) raised in the topic sentence. When it is time for a new topic sentence, it is time for a new paragraph.[11]

---

[9] Having learned to do so more quickly than in the quote attributed to Oscar Wilde, that he spent an entire morning taking out a comma and an entire afternoon putting it back in.

[10] This paragraph's invisible header is "What is a paragraph?"

[11] This paragraph's invisible header might be "The use of topic sentences and the structure of paragraphs."

So, we can consider a hierarchy of audible pauses or pauses in thought for the reader that a writer can use to pace and control the rhythm of a piece of text. These are, in order of increasing brevity: hyphen, comma, semicolon, full stop, and paragraph break (and perhaps beyond that section break with new topic headers).

PRINCIPLE 4: LOOK AFTER YOUR COLONS

Your colon is an important part of your body; similarly, a colon, and its more powerful cousin, the semicolon, are important parts of scientific rhetoric. Colons and semicolons break up sentences where a full stop would not be elegant, and allow the equivalent of dramatic pauses and showman-like revelations. These items allow a little implicit excitement to creep into scientific writing, and should be part of the writer's toolkit, albeit ones to be used cautiously and correctly.

A colon is usually followed by something that is not a sentence: a piece of information, fact, or data point. In most cases, there is another (maybe marginally more word-requiring) way to avoid a colon, and in such cases the second route should be taken, for the colon-based version seems unnecessarily pretentious and formal.

"I would like to thank the following for their help: Jim and Mary."
"I would like to thank Jim and Mary for their help."

So, I tend to avoid colons in my scientific writing, and excise them routinely as editor.

I am, on the other hand, a big fan of semicolons.

A semicolon is followed by something that could be a separate sentence, but is so intrinsically linked to the sentence-like construction that precedes the semicolon that it makes explanatory sense to present both in such a conjoined manner.

In science, that which follows the semicolon typically explains or comments on that which precedes it, and provides context and commentary in a very concise manner.

The kinetic data (Fig. 3) indicate that chemical reaction proceeded at a faster rate than predicted; this may be due to an unanticipated impact of the presence of compound XYZ, which acted as a catalyst.

There is a certain grace in a well-used semicolon, but they should not be excessively used. Indeed, some writers have expressed strong disdain for them, with Kurt Vonnegut once saying that "all they do is show you've been to college" (presumably, he wouldn't argue with scientists using them, then), although George Bernard Shaw once admonished T. E. Lawrence for practically not using them at all, which Shaw deemed a "symptom of mental defectiveness." Clearly, semicolons inspire heated views on both sides!

I recently received a referee's report on a student's paper which commented sarcastically that we had an overreliance on the use of the semicolon. As a result of that comment, I actually searched the book manuscript I had almost finished at that time for that grammatical symbol and probably removed 75% of them. I may have been overenthusiastic in reflection, but it is never too late to take and act on constructive feedback; no writer should be so arrogant as to think they know it all and can never improve.

The one place where I condone the use of colons is in conjunction with semicolons in constructing a list, where the elements of that list are themselves complex and convoluted, such that separating by commas is not sufficient to ensure that the individual items on the list are seen as indeed individual elements (especially where there are *and*s in more places than just before the final item). This is essentially a polite way of doing a bullet point list (where the semicolons are the bullets) in circumstances where bullets would look inelegant or out of place.

> The key characteristics of a good scientific writer are: a good understanding of grammar and syntax; an ability to take and learn from constructive criticism; and strong attention to detail in presentation, spelling, and formatting of text.[12]

PRINCIPLE 5: KEEP IT SHORT, IF NOT SWEET

If clarity is the first virtue of good writing, brevity follows close behind. Scientific writing should use exactly the minimum number of words to make a point clearly, and not one word more. One of the most common misjudgments of the inexperienced writer is to use unnecessary words, littering a manuscript like, well, litter.

In my experience, when people first attempt scientific writing, they have a natural hesitancy, and preface what they want to say with obfuscation and delay-words (I know I did, when I look back at early examples of my projects and theses).

Consider the following example:

> It can be seen from the results that the action of the compound reduced inflammation in the lungs.
> The results show that the action of the compound reduced inflammation in the lungs.

In both cases, multiple unnecessary words delay getting to the core point, which begins, "The action . . . ."

Another common long-winded construct says "resulted in a decrease," which is exactly the same, if longer, as saying "decreased," while a particular bugbear is "affected/changed/altered," where this word can be replaced by "increased/

---

[12] I must admit that the "and" after the final semicolon, which appears to be technically correct, annoys me, but both versions with or without it look equally clunky to me, even if it was moved before the semicolon.

deceased/other direct verb," clarifying information that otherwise needs to be provided in an unnecessary bunch of extra words.

When editing text, I spend a lot of time watching for sentence flow and length, and have an intuitive feel for when sentences are too short, which is when they render the text very "start-stop" in feel and staccato, or too long and convoluted. I wouldn't recommend any formulation as to the ideal length of a sentence, but if the attention wanders while reading it then it needs total (full stop) or partial (semicolon) respite points.

PRINCIPLE 6: MIND YOUR HYPHENS

The hyphen is the anticomma, and saves you words if used right. One good way to stick to principle 5 is to use the hyphen. Hyphens allow us to put words together in specific ways and specific orders in a sort of short-hand while making sure the intended meaning does not get mangled; one exception is where the first word being glued to a second ends in "ly" as this clearly already links to that which follows immediately after.[13]

For example, we could say "samples treated with heat," or "heat-treated samples" (saving one word, and every one counts!). Hyphens also allow us to clarify the way in which words are linked. For example, "disease causing bacteria" suggests a return to old-fashioned notions of spontaneous generation, while "disease-causing bacteria" makes the point without the reader having to think twice or reread the words, which should always be the case. Inserting a hyphen can also completely reverse the meaning of a few short words, as in one of my favorite examples, "bacteria carrying dust particles." Likewise, consider an "abnormal occurrence report"—are the occurrences abnormal or is the report?

I call the hyphen the anticomma because a comma in text suggests you take a pause between words, while a hyphen tells the reader to read the words quickly together. In speech, if you run words together like this, that's where the hyphen goes on the page. To consider the complexity of the hyphen, and its power, a little further, consider "treatment of samples at high temperatures," "high-temperature treatment of samples" and "high-temperature-treated samples"; all are correct uses, showing how a phrase can be rearranged using hyphens to ensure the order and sense don't get lost. In the last case I used a double hyphen, which isn't seen

---

[13] There is a wonderful example of where to use and not to use hyphens in Bill Bryson's book *The Road to Little Dribbling*, which is worth quoting in its entirety: "And yet with a few unassuming natural endowments, a great deal of time, and an unfailing instinct for improvement, the makers of Britain created the most superlatively park-like landscapes, the most orderly cities, the handsomest provincial towns, the jauntiest seaside resorts, the stateliest homes, the most dreamily-spired, cathedral-rich, castle-strewn, abbey-bedecked, folly-scattered, green-wooded, winding-laned, sheep-dotted, plumply hedgerowed, well-tended, sublimely decorated 88,386 square miles the world has ever known—almost none of it undertaken with aesthetics."

too often but is perfectly valid and can help marshal awkward phrases into orderly sensible queues of words, the ultimate destination of which is clear.

PRINCIPLE 7: NEVER EVER MISUSE
AN APOSTROPHE

There should be no apostrophes in simple plurals, ever—nonscientists may break this rule today with jaw-dropping alacrity (CD's, 1980's, birthday's, etc., ad true nauseam), but if you misuse them as a scientist, you should have your PhD revoked instantly. Apostrophes indicate either contraction[14] or possession, not plurality.

PRINCIPLE 8: HAVE A GOOD VOCABULARY, BUT
DON'T SHOW IT OFF

It is important and inevitable that a good writer will have an extensive vocabulary, with an extensive arsenal of choices of word for every occasion. However, as with many arsenals, having them doesn't mean you have to use them excessively, and it is always better to have and not need then to need and not have. I got quite giddy recently when I came across a word in an article in the *Guardian* newspaper that I had never heard of: contumaciously.[15] I checked the meaning and put it on a shelf in the mind-palace of writing tools in my head, while being pretty sure I would never find occasion to use it ever again (outside this page, and perhaps a very weird party).

Never should a complex or show-off word be used where a more familiar one will do the trick. For example, you should avoid saying "prior to" or "subsequent to" unless it is before or after you want to appear like you have swallowed a thesaurus and are burping it up one word at a time.

PRINCIPLE 9: DON'T GET BLINDED
BY THE SCIENCE

This may seem counterintuitive, but the problems with scientific writing often relate not to the complex scientific terminology used but to the smaller words and grammatical constructions that form a scaffold around it. A further key point is that the brow-furrowing effort on the part of the reader should come from seeking to understand and analyze the ideas and concepts being discussed, not the way in which they are presented. Sometimes a piece of scientific text contains an intimidating array of long terms or names, acronyms, and polyhyphenated constructs, all of which draw the eye and obscure the sense of the sentences in which they are embedded.

---

[14] As in "isn't" for is not, a shorthand that I find too informal for most scientific contexts.
[15] Which apparently means stubbornly persistent or refusing to recognize authority or the law.

One way to simplify writing when dealing with these juggernauts of jargon is to eliminate them entirely, by replacing them with a simple letter code (where I write "A1" I mean "[specific long technical term]"), and then writing sentences free of their looming shadows. Then, when the text has been polished into coherent fluid readability, the codes can be broken and the original terms reinserted, hopefully settling neatly into sentences that have been sufficiently reinforced not to creak or snap under their weight.

In a small but related point, sentences should start with a word with a capital letter, not an acronym, symbol or digit, and so we might say "One litre of buffer" rather than "1L of buffer." We also need, as always, to consider our audience, but assume that all acronyms and abbreviations used will need to be explained at first point of use, and not thereafter.

Having got these propositions (and a couple of prepositions) off my chest, let's now turn to the broader assembly of written scientific prose, and start with where the modern style originated.

The Historical Development of Scientific Writing Style

In reading classic scientific papers, it is hard not to be struck by how the modern style of impersonal prose was not always the norm.[16] Many papers are written as if they were direct transcripts of lectures, in a declamatory style, which may reflect their origins. For example, Max Planck's 1900 paper on the distribution of radiation spectrum, in *Verhandlungen der Deutschen Physikalischen Gesellschaft*, begins, "Gentlemen: when some weeks ago I had the honour to draw your attention to . . . ," and includes linking phrases like "I shall now make a few short remarks about . . . ."

The paper by William Bayliss and Ernest Starling on the mechanism of pancreatic secretion (essentially the discovery of hormones) in the *Journal of Physiology* in 1902 reads almost like a thriller, with a section heading "The Crucial Experiment" that includes the words "But, and this is the most important point of the experiment, and the turning-point of the whole research . . . ."

While I commented earlier how common it was in former decades and centuries for reports of major discoveries to have a single author, it remains notable how many of those were written with usage of what might be called the "royal we," where an author uses that pronoun rather than the more obvious "I" in their writing.

A certain gentlemanly relationship between scientists, who in many cases were probably rivals, also shines through many classic papers, as when Niels Bohr in his paper "On the Constitutions of Atoms and Molecules" in the *Philosophical*

---

[16] Many of the examples that follow were facilitated by a lovely book called *The Discoveries* by Alan Lightman (New York: Pantheon Books, 2005), in which excerpts and critiques of the original papers are collated.

Magazine (1913) referred to "Prof [Ernest] Rutherford" right at the start of his paper, and finished the introduction by stating that "I wish here to express my thanks to Prof. Rutherford for his kind and encouraging interest in this work."

In their paper on the existence of alkaline earth metals resulting from neutron bombardment of uranium (a key paper in nuclear physics), Otto Hahn and Fritz Strassman include a number of unusual expressions and stylistic choices, including an exclamation mark after one of their observations on the sum of masses of elements, and a note of modest and self-effacing caution rarely seen in the scientific literature as follows:

> As chemists we really ought to revise the decay scheme given above. . . . However, as "nuclear chemists" working very close to the field of physics, we cannot bring ourselves yet to take such a drastic step which goes against all previous experience in nuclear physics. There could perhaps be a series of unusual coincidences which has given us false indications.

Finally, while language has probably become more formal and less personal over time, scientists throughout have sought to keep certain distance from language that might be seen as simply too unscientific. For example, Arno Penzias and Robert Wilson, when discussing the way in which they eliminated alternative sources of the noise in the large exquisitely sensitive radio detector they were using, leading to the Nobel Prize–winning conclusion that that noise represented leftover electromagnetic radiation from the Big Bang (which must therefore have happened), referred to clearing "white dielectric material"[17] from the antenna.

The wording of the 1953 *Nature* paper by Watson and Crick on the structure of DNA has been particularly widely examined (for many reasons, some touched on elsewhere in this book) but the confidence that characterized both individuals leaps off the page in wording like that used after they have swiftly demolished all previous proposals for the structure of that molecule: "we wish to put forward a radically different structure."

Today, a notable characteristic of most scientific writing is the pursuit of objectivity, to the point of removal of all traces of emotion or, arguably, personality, with the presumed objective of presenting the work in the most dispassionate and evidence-based manner possible. This might be most readily evident from the typical current absence of personal pronouns such as "I" or "we" in many papers. This results in some rare cases where more words are used than are strictly necessary, or a more complex construct than strictly necessary, to avoid those damned pronouns; instead of saying "We measured the pH of the samples," we find "the pH of the samples was measured."

Nonetheless, fragments of emotion can come through the process, as when the first paper on the structure of Buckminsterfullerene (the 60-carbon spheres shaped

---

[17] Bird (specifically pigeon) poop.

like soccer balls) referred to the model as "an unusually beautiful (and probably unique) choice."[18]

Some scientists fight the system a little, to allow some levity into the austere sincerity of scientific literature. There is a story about a bet in the 1970s between some researchers over a game of darts, which meant that the loser had to use the word "penguin" in their next paper. This led to a particular type of particle physics diagram being described in a rather contrived manner, and presumably rather unconventionally, as a penguin, by which name it is actually still referred to today. In a similar example, a group of Swedish physiologists have for years been sneaking Bob Dylan lyrics into a number of papers (an example being a *Nature* paper titled "Nitric Oxide and Inflammation: The Answer Is Blowing in the Wind").

A number of wedding proposals have also apparently been incorporated into papers, including a recent physics paper that was subscriber-only,[19] leading to a tweet that romance was not dead, just behind a paywall, and an even more ambitious example where a physicist wrote an entire (fake) paper on the progress of their relationship ("Two Body Interactions: A Longitudinal Study"), which concluded with a proposal of indefinite continuation of the study (i.e., marriage) and a tick-box yes/no option for completion. Both cases mentioned apparently led to a happy ending!

Papers can also include hidden political comments—it was reported in late 2018 that a paper had been withdrawn from the journal *Scientific Reports*, because the authors had chosen to illustrate the topic of recovery of DNA from fecal material with a diagram in which, on very close inspection, a certain then-current US president's face could be detected as being inserted into a sample of the aforementioned material.

In 1971, a paper was published in the *Journal of Organic Chemistry* in iambic pentameter, but this sadly failed to start a new trend, while I very much enjoyed the following paragraph, which appeared in a paper in *Science Advances* in 2016:

> The particles in this article (Although "the particles in this article" is in this particular article, consider "the particles in an article" as part of an article. As any articulate party would know, the particles in "the particles in an article" are "the" and "in," whereas the articles in "the particles in an article" are "the" and "an," but the particular article in "the particles in an article" is "the." "p.s." is all that is left when you take the "article" out of "particles.") are photons, as was the case in Kocsis et al.

---

[18] The authors (Kroto et al.) also refer to being "disturbed by the number of letters and syllables in the rather fanciful but highly appropriate name we have chosen in the title to refer to this . . . species . . . a number of alternatives come to mind (for example, ballene, spherene, soccerene, carbosoccer), but we prefer to let the issue of nomenclature be settled by consensus." Consensus settled on buckyballs.

[19] "Rui Long wants to thank, in particular, the patience, care and support from Panpan Mao over the passed years. Will you marry me?"

A final area where scientists could exhibit personality is in naming their findings, such as a pungent organic ring-shaped chemical called arsacyclopentadine (arsole for short) and a gene intriguingly called Sonic Hedgehog.[20] This tendency to humanize the complex through humor has manifested through campaigns to name scientific entities things like Boaty McBoatface (an Antarctic exploration vessel) or lemmium (a newly discovered heavy metal, proposed—as in the case of the ship, unsuccessfully—to be named after the late singer of the band Mötorhead, noted proponents of that particular type of music).

Rhetoric and the Construction of Sound Arguments

Having established the need for a scientific paper to be (relatively) emotion-free and objective, it still has to have impact. It needs to explain, persuade, defend, attack, and above all convince.

The key objective of a scientific paper after all must be to present a convincing argument for why a piece of research needed to be done, that it was done in the best way possible, and that the results as explained by the authors are convincing and important. Every paper must be an argument, backed up by reference, statistics, data, and language. The discipline of constructing logical arguments is rhetoric, and, while scientific writing is not often discussed in terms of rhetorical tools, this topic deserves a brief mention here.

The Greek philosopher Aristotle articulated three ways to appeal persuasively to an audience, which were logos, pathos, and ethos. Logos represents the logic behind an argument, including the use of supporting evidence, and is a key principle in scientific communication, where statements must be backed up, usually by reference to a preceding source or new evidence. Pathos refers to appealing to the emotions of an audience, which perhaps relates more to the tools that a researcher will use when making a presentation (as we will discuss in chapters 8 and 9), and ethos to the guiding principles and ideals of a community, which links to concepts of standards of behavior and integrity that we will discuss in chapter 7, but also relates to the credibility of a speaker or writer, which underpins all forms of scientific communication and how it is received.

The Roman philosopher Cicero laid out five canons of rhetoric, as follows (with modern interpretation):

Invention—the consideration and brainstorming of a piece of writing or talk, or developing and refining arguments;

Arrangement—the construction of text around a structure and order of points so as to achieve maximum impact;

---

[20] The fruitfly, a famous test organism for genetics, has, besides Sonic Hedgehog, genes named Daschund, Swiss Cheese, Van Gogh, Ken and Barbie, and Indy (I'm Not Dead Yet).

Style—the use of language to present effective arguments;

Memory—committing material to memory, perhaps when a researcher prepares a presentation by rehearsal;

Delivery—particularly when presenting to an audience.

So, while researchers might not think on a frequent basis about the influence of Greek and Roman philosophers on what they write or how they prepare for a talk, principles that were laid down thousands of years ago actually describe many of the key actions and approaches they will take. Science is about arguments, and a little knowledge of rhetoric can show a lot about how arguments are best approached, and won.

The functions and goals of a scientific paper can be achieved with the assistance of the correct deployment of language and rhetoric. The points to be made are identified and laid out in an implicitly rhetorical structure, and then their deployment proceeds through careful use of language and grammar.

The objective flows from the statement of need in the Introduction of a paper through suitable linking phrases to draw the reader from the gap in knowledge to the existential reason for undertaking the research in question (the Invention): "therefore, to address this issue, the current study was designed to . . . ." The methodology is then presented using its own precise formulations of instruction, followed by the results laid out with their accompanying narrative.

Then, in the discussion element of the paper, language must serve the functions of Arrangement and Style to convince the readers of the correctness and importance of the findings, as always seeking at every possible opportunity to stress what is new, what is different, and what changes our view of the topic in question: "this is the first report of . . . in contrast to the findings of . . . it can clearly be seen for the first time that . . . ."

A researcher can spend weeks, months, or years on a piece of work but, when it comes to presenting their work formally to the world, they are curiously absent bar their name, and there is no explicit sense of the time invested. It is all distilled into a couple of thousand words, to be read in perhaps an hour or less, but it is expected to have a longer-lasting impact on the readers. Thus, language must be the researcher's vehicle to communicate their message as, between their efforts and findings and the impact these might have is found the gateway of written communication, and if this is not traversed as effectively as possible the desired impact may never be achieved.

The Importance of the Right Writing Environment

Having considered the grammatical toolkit a good scientific writer needs, and the style they should recognize as generally expected, let's turn to practical considerations. When preparing to write, whether a draft of a paper, chapter, thesis, or other document, certain things need to be in place: writing tools, time, and a writing place and environment.

Let's consider each of these in turn.

Tools we will not delay on excessively, presuming we are discussing in most cases a computer or laptop with appropriate word-processing software. I would mention that, as an experiment, I wrote some of this book using Google Docs, and loved being able to access the most current version from my phone, iPad, laptop, or home or university office computer instantly, to do a chunk of writing or just make an occasional note or edit when it popped into my head.

I would note though that, if like me you might occasionally want to go backward to an earlier draft to check something or revert to an earlier version of something, you need to adopt good discipline for regular saving of dated drafts offline. Indeed, one of the skills any researcher needs to master early on is that of effective document management, particularly in terms of tracking drafts, feedback from coauthors, et cetera.

This raises once again the point that very few scientific documents are completely sole-authored, and are likely to go through multiple drafts and revisions as others add their comments, sections, and questions into the mix. However, in most cases, someone has to take responsibility for creating a first draft that will form the basis for such revisions, and, as this is frequently the most junior member of the team, this will be the scenario considered here.

Let's next consider time for writing. Hugh Kearns of Flinders University in Australia, who runs workshops worldwide to researchers on the subject of overcoming procrastination in writing (among other topics) speaks of the importance of two golden hours, as a sort of optimal block of time to allocate at one go for writing, being long enough to allow establishment of a rhythm, concentration, and engagement with the material to be achieved, while not being so long that these start to wear off. In those two hours, notably, the task is writing, and it should not be compromised by phones, email, social media, or other distractions.

The key point is that focused writing is not something that can be done in snack-like periods of twenty snatched minutes here and there, rushing between things or waiting for a measurement to be complete. Nor does it work well in a busy environment, where someone might have a couple of hours but be regularly interrupted in one way or another throughout by demands that absorb both time and concentration, and can cause lingering distractions that destroy any possibility of a stream of effective thought.

When giving workshops, I used to say this was critical for those starting to write to give themselves the best chance of productivity by getting such blocks of time. However, as I started in recent years to work on major book projects like this I found that this was absolutely the best model (2- to 3-hour blocks at a go), both to set aside dedicated time in the calendar and to get into an appropriate groove in which I felt I could make genuine progress (although I can't admit to being sufficiently disciplined to ruthlessly block out calls, emails, and such like during that time).

Having thus carved out some time to write, the final requirement for successful writing is a suitable location and environment (not the same thing). In terms of location, the typical scientific workspace of a laboratory is often highly unsuitable, due to noise, action, and other individuals being present, as well as possible issues like smells and safety concerns.

For those lucky enough to have one, an office (home or work) with a suitable surface, electricity, comfortable chair, and a door that can be closed, probably with decent wi-fi (and sometimes, when concentration is needed, the absence thereof), is optimal. This is where the difference between location and environment comes in as, for some people this sanctum needs to be quiet to facilitate concentration, whereas for others (myself definitely included) music is a good accompaniment, not too distracting or loud but which makes me relaxed and which I enjoy (probably not something I haven't heard before), and definitely not anything that might require attention, such as podcasts, news, or audiobooks.

Most of this book was written in my home office, some in my garden (including these very paragraphs, on a rare warm Irish early summer's day), some on trains and planes (although I prefer those for editing rather than creative fresh writing), and some in cafes or public spaces. There is a certain theory that cafe noise is optimal not-quite-white-but-featureless-yet-sociable sonic background for writing and, to facilitate this, there is even a website that can pipe a range of variants of such noise into your writing environment (coffitivity.com).

As with so many aspects of writing, different techniques will work for different people and it may take some time to find the Pooh-esque "thotful spot" that has the right vibe for effective writing for that individual, and for them to then structure their time to find opportunities to retreat there and work. But what happens next?

Writing Styles and Habits

Having found a good spot, with the right noise, light, solitude/company, and having carved out two sacred hours in which to write, the writer then opens their word-processing system of choice and is confronted with the simultaneous terror and promise of the empty page.

Here again individuals will diverge in preference and recipe for success. For some, the best approach is to complete one paragraph, fully checked and referenced, before moving to the next paragraph, and so forth. Others will be like architects who map out the structure of their document or paper in extensive detail, section and subsection by subsection, before starting to build in text around the framework thus erected.

For yet others, myself included, the key initially is to get text on a page in a form of free-writing which is almost a stream of consciousness, in which key thoughts, ideas, and notes are dumped onto the page, in what might initially be a ramshackle fashion, but which is then progressively wrangled into shape across later drafts as it is rewritten. In this model, details like references might not be added until perhaps the third pass, as the document slowly takes shape layer by layer, like a painting.[21]

When working on my first long (book) project (*Molecules, Microbes and Meals*) I found that, in the early stages, for each block of time I had, I literally spent the first few minutes wondering what I felt like writing about at that time (working from a chapter-based skeleton, in most cases subdivided into a few indicative section headers), and then got stuck in there. It didn't matter if that was before or after, close to or far from, that which I had written in the previous writing session. This odd approach of course meant that I kept putting off certain sections which eventually had to be tackled, but by the time I got to them I had made a lot of progress on the rest and so was on a bit of a roll. It also meant that I had to spend quite some time later knitting these disparate text fragments into (hopefully) a coherent whole, but that was mostly a fun exercise, which even involved text sections occasionally leaping from their original intended home to a whole new place, chapters swapping places, or bits coalescing to form a new and originally unexpected chapter.

These developments flowed directly from the document being in my mind a set of almost independently mobile elements that could be rearranged in a wide range of ways. This was not quite as chaotic as it may sound, because of the constraints of the first proposed structure, but the approach was flexible enough to allow that structure to be deconstructed and reconstructed as the project evolved. All I can say is, it (hopefully) worked![22]

Writers will also differ in their speed of writing, and in how long it takes to complete an assignment. Again, some will like to write carefully and perfect each point before moving on, while for others (myself included) a preferred style is to write fast and edit and fix later, once the raw material is in place to work with. Of course, when writing, as with all other aspects of science, quality is more important than quantity, as in the old joke about someone not having time to write a person a short letter so they wrote a long one instead. I once saw a lovely pencil with a word count along the side which, in theory, reflected the position which the pencil parings had brought it to with ongoing writing—much more charming than word counts on a computer.

In a final comment on this topic, I am reminded of a joke about the writing habits of James Joyce, who apparently was famously slow in writing. A friend once came to his lodgings at the end of a day and found him slumped over his typewriter in despair: "I only got five words written today," he moaned and, when his friend

---

[21] Having dabbled from time to time myself with applying some modest ability in painting, the analogy of building up text in layers like oil paint on canvas repeatedly occurs to me when considering how I approach writing.

[22] By comparison, this book was written in a far more linear fashion, as the chapters represent reasonably independent topics, ring-fenced from their neighbors, but the order in which these were written is certainly not that in which they appear, and streams of consciousness and free-writing played a key role, including one feverish summer's day toward the end of my first draft when I discovered I had written 8,000 words, leaving me in need of a rest and the text in need of a good edit.

said that this was actually pretty good going for Joyce, the writer responded "yes, but I don't know what order they go in."[23]

Collaborative Writing

It must of course be acknowledged once again that very little scientific writing is undertaken as a wholly individual activity. The mean number of authors per paper is steadily rising and, even for someone not in a hyperauthorship area like particle physics, it is likely that many researchers will have at least two, and maybe typically up to five, coauthors.

These may be divided into different categories:

1. The genuine coauthor—a researcher who has a critical direct writing input, describing and discussing part of the work and results, and in this way contributing a significant proportion of the final article;
2. The supervising coauthor, who has overseen the work and been involved throughout, but is more likely to edit and give feedback than write actual text;
3. The sleeping coauthor, whose contribution is less than those above, but has somehow met the criteria for authorship sufficiently to be named. Care has to be taken that those who have not been directly involved in the work and/or writing are not in the category of honorary author, and all authors should still meet the criteria for authorship such as those articulated in the Vancouver Protocols (discussed in chapter 3).

No matter how many authors there may be, they are all supposed to input into and sign off on any paper on which their name appears (and so be equally prepared to share for the paper in question both credit and advantage, in the best case, and blame and responsibility, in the worst case). To manage this may become a complex exercise, as comments and feedback (sometimes mutually contradictory) are received from multiple sources, and the lead author needs to have a good work plan in place for managing such inputs. Elements of this will probably include a clear timeline (and agreed order, as generally in my experience this works best if managed in sequence, rather than parallel, with multiple authors working on the same draft) for who receives drafts when, and by when they are supposed to return feedback. There also needs to be agreement on the form in which feedback will be supplied, and an effective system for labeling and tracking drafts.

Of course, the process will become proportionately more complex the more authors there are, and it seems highly unlikely that papers with massive author lists

---

[23] I am sure there is a secondary joke in there about whether he ever worked out the answer to that one, but I can't admit to having the prerequisite experience of having managed to read the books in question to be allowed to make that one.

have been read word-for-word by every author before publication (and definitely not in sequence!).

The sociologist Harry Collins published in his book *Gravity's Kiss* a lovely account of how this worked for one of the most high-profile physics discoveries in recent years (a crowded field, given the Higgs Boson and whatnot), which is the discovery of gravitational waves.

Gravitational waves were first predicted by Einstein in 1916 to explain how large bodies in space transmit the effects of gravity, through their mass creating distortions in space-time, that keep them in orbital or otherwise linked motion. Without gravitational waves, how would the earth "feel" the pull of the sun, or the moon that of the earth? The problem in confirming this prediction lay in the extreme difficulty in detection of what was estimated to be a very weak (although, in other ways, really powerful) force. Detectors designed to capture them were developed based on measuring astonishingly small changes in the passage of light beams, which were believed to be shaken, ever so slightly, by one of these waves happening to pass by, like a boat moving a tiny amount as a ripple passes through the pond beneath it.

In 2015, one such detector system (called LIGO, the Laser Interferometry Gravitational Observatory) detected a signal that was consistent with one of these waves infinitesimally nudging one of its laser beams, and careful analysis over a period of months led to a conclusion of 99.9994% certainty (physicists need a high degree of confidence for their biggest claims!) that the result indicated the detection of gravitational waves, and also that the waves originated from the merger of two black holes, each around 30–40 times the mass of the sun, 1.4 billion light years or so from earth (some days I really wish I were a physicist).

Gravity's Kiss is an account of this discovery and the writing of the resulting paper,[24] which included just over 1,000 authors, in 133 institutions, three of whom are listed as deceased in footnotes. How can a paper this important be written by so many people? Apparently, the answer involved the formation of a drafting committee (with subgroups for specific parts like the abstract), a mind-boggling number of group emails, many huge teleconference calls (with up to 290 centers connecting at once) and, on a number of key points, an online ballot to get consensus on specific points for the paper and how they were to be written. Disagreements ensued throughout the drafting about the directness of the wording (such as the difference between in effect saying "we saw" and "we think we saw"), the extent to which historical background would be referenced, and whether the observation could correctly be described as "direct" or not. Ballots offered all 1,000 authors a chance (which 288 seized) to rank in terms of preference eight different titles and four different options on the use of the word "direct," and the final versions in the published paper arose from such painstaking diplomacy.

Concerns occasionally emerged about whether the growing draft showed too much evidence of having been, literally, written by committee, and therefore

---

[24] Abbott, B. P., et al. (2016) Observation of gravitational waves from a binary black hole merger. *Physics Review Letters*, 161102.

lacked a single commanding voice, but eventually the final draft was circulated to all authors for a vote on whether it was ready to submit. Only 5 out of 592 voters thought it was not, and so it was submitted, received very positive referee reports (within a week) and was published shortly after, creating waves in physics of far more noticeable magnitude than those detected from those distant black holes.

The Central Importance of Clarity

Above all else, a paper is useless if it cannot be understood by its target audience. Writing in a high-falutin' style that may seem impressively baroque and fancy is in no one's interest, and this leads back to my points made earlier about the best scientific writing being that which is most functionally effective in transmitting its message without fuss or excessive challenge on the readers to understand the writing (whatever the contents).

One function of the review and editing process should be to screen papers for their clarity and accessibility, but there is one somewhat extreme example that could be interpreted as showing the way in which opaque text can be interpreted as depth.

This was a paper titled "Transgressing the Boundaries: Towards a Transformative Hermeneutics of Quantum Gravity," which was published in a journal called *Social Text* by a physicist called Alan Sokal. Publication of this paper was followed quite quickly by a second paper (in a different journal) by Sokal titled "A Physicist Experiments with Cultural Studies," in which he admitted the first paper was a hoax, designed to test whether a sufficiently pretentious-sounding paper could really be published without proper review by experts in the field (of course, it should be noted that he submitted the paper to a journal outside the area of physics, which should arguably have rejected it as out of scope).[25] The paper is full of parodies of over-wrought language,[26] jokes,[27] deliberately ridiculous and provocative statements,[28] and statements intended to flatter the editors of the journal to which it was submitted.

---

[25] The entire incident and its background and aftermath, as well as an annotated version of the paper explaining all the jokes and references, are presented by Sokal himself in his book *Beyond the Hoax* (New York: Oxford University Press, 2008).

[26] Such as "the discourse of the scientific community, for all its undeniable value, cannot assert a privileged epistemological status with respect to the counter-hegemonic narratives emanating from dissident or marginalized communities"—eh, what?

[27] Apparently, the line "liberal (and even some socialist) mathematicians are often content to work within the hegemonic Zermelo-Fraenkel framework (which, reflecting its nineteenth-century liberal origins, already incorporates the axiom of equality) supplemented only by the axiom of choice" plays with multiple mathematical and cultural meanings of the words "equality" and "choice." I am sure those smarter than me are ROTFL at that one.

[28] "The content and methodology of postmodern science thus provides powerful intellectual support for the progressive political project, understood in its broadest sense: the transgressing of boundaries, the breaking down of barriers, the radical democratization of all aspects of social, economic, political and cultural life" (clearly).

This is (obviously) an extreme, yet I have read many papers that don't rise to this level of magnificent absurdity but remain very difficult or tedious to read. The Canadian philosopher Marshall McLuhan coined the phrase "the medium is the message," meaning that there is a symbiotic relationship between the medium by which a message is conveyed and the message itself, and that the medium influences how the message is perceived. In scientific publishing, the medium on one level is the journal in which work is published (which certainly influences how that work is perceived) but at a more fundamental level the medium is language, which is the medium through which the message of science must be passed. If not deployed well, language can become a barrier to comprehension, interest, and ultimately uptake and impact.

5

The Perils of Peer Review

One small step for an author, one giant leap for their work:
submitting the paper

Usually, the principal outcome of a piece of research that will be expected of researchers, by others and themselves, will be publication of a paper in a respected peer-reviewed journal, which will be the focus of this chapter.

It has been said of paintings and movies that they are never finished, only abandoned. This can also be applied in many ways to scientific papers. The molecular biologist and Nobel Prize-winner David Baltimore once said

> No study is ever complete. . . . Deciding when to write up a study is an arbitrary and personal decision. A paper is written when an investigator decides that a story can be told that hangs together, that makes sense and that others will want to read and build on. The scientific literature is a conversation between scientists. . . . It is crucial to remember, and often forgotten, that a paper does not claim to be an absolute assurance of truth, only a moment's best guess by a group of investigators. Because all of these judgements are less than wholly objective, another investigator might have come to a different conclusion using the same data.[1]

At some stage, the paper is finished, or at least abandoned, and the journal to which it will be submitted must be selected. This point may come before the paper is written, midway during its gestation, or when it is complete, and may be, as with all aspects of publication, a source of disagreement among authors. The selection of the journal, including whether the publisher is traditional or open-access, will be on the basis of factors discussed in chapter 2.

Based on factors such as these, a journal of first preference will be agreed, as well as perhaps a backup list of next preferences. The next stage is to check the instructions for authors (to be found on the journals' website or within each hard-copy issue) of that journal, to ensure that the paper manuscript conforms to these

---

[1] Kevles, Daniel J. *The Baltimore Case*. (New York: W.W. Norton and Co., 1998), p387.

instructions in terms of referencing style, length, presentation, and so forth. Every journal has its own preferences, and editors of journals get annoyed when people ignore these instructions. Such transgressors accumulate black marks against them from the start, and simple professional courtesy demands that guidelines provided be adhered to. Selecting the target journal early in writing can ensure that the paper is written with the required style in mind, as opposed to this being superimposed at the final stages.

Authors, as well as following the journal's presentation guidelines, typically prepare their manuscripts in double-spaced format, with line and page numbers (to help referees draw attention to specific parts of the paper in their reviews), and tables and figures presented at the end of the paper, on separate pages (with figure legends also presented in a separate list). In some cases today, particularly for open-access journals with rapid ideal publication speeds, authors may be requested to present their manuscripts in a format that is close to the final appearance in the journal.

The instructions for authors also provide practical instructions on how to submit the paper to the journal. A chief editor, selection of editors, or editorial office will also be identified in such places.

When I started publishing papers in the 1990s, submission entailed sending three hard copies of the paper, paper-clipped together, to the appointed person, along with a cover letter requesting politely that the enclosed paper be considered for publication. Today, as with so many aspects of science, an electronic revolution has taken place. For most (if not all) journals, hard-copy postal submission is a thing of the past.

What succeeded snail mail, however, evolved rapidly in a short space of time. In the late 1990s, for many journals, the practice became to email manuscript files to the editor or editorial office; this saved huge amounts of time involved in files traveling between editors, authors, and reviewers, who were frequently geographically scattered around the world.

However, this was quickly replaced for most large professional publishers by online submission and editorial platforms, for which all parties (editors, reviewers, authors) would register to set up accounts to manage their various roles in the process. Each large publisher now has their own version of such a platform, within which each of their journals has a separate subplatform, with similar basic functionality but differentiation in terms of logos, color schemes, and so forth. These systems are then used by authors for uploading their manuscripts and associated information, by editors to find and work with reviewers and authors, and by reviewers to access agreed manuscript assignments and complete their review reports.

Submitting a paper to such a system is a relatively straightforward, if somewhat tedious, process. Having logged into a new or existing author account, the individual responsible for submitting the paper (probably the corresponding or first author) will work through a succession of screens in which they will probably be asked to enter the paper's title, abstract, and author list (and contact details), as well as, depending on the journal, things like suggestions for reviewers (or, occasionally, suggestions of who the authors do NOT wish to review their paper), evidence of ethical approval if needed for the research described, original signatures of all authors named on the paper, and affirmation that the authors are aware of, and not in infringement of, the journal's policies on research integrity or plagiarism.

Having worked through these questions on successive screens, the submitting author will eventually get to a screen where they are asked to upload the files containing their paper, typically the manuscript plus separate files containing tables, figures, figure legends, any additional required files such as original data, and any supplementary material. Most journals will also expect a cover letter, which is typically short and to the point:

Dear [Editor]

We would like to submit the accompanying paper titled [XX, by YY] for consideration in [Journal name], and confirm that this paper is not currently under consideration elsewhere.

Yours sincerely,

The cover note is not the place to make an impassioned argument for the merits of a paper (the paper should do that, and the referees do not see the cover note) but it may be the place where any additional information of relevance to the editor is presented (e.g., if the paper is a new version of a paper previously rejected by the journal or there is some other relevant history).

Finally, when the author confirms they have uploaded all the relevant information, the system will generally take a few minutes to churn through and check the files, and then present a PDF file into which all the documents have been assembled, as a proof of what the system believes the authors wish to submit. This is the last chance for authors to check that all figures are correctly included, no pages are missing, and that the authors really want to subject their work to the process of peer-review with that journal. They then, if they are happy to proceed, click a link called "Approve submission" or such like, typically with a small amount of trepidation, as that is the last action they will likely take until peer review is complete.

Once that link is clicked, the authors have completed the first critical part of their duties; however, the work is only beginning for others as, immediately after that click, somewhere in the world, someone (probably the journal editor) will get an email or notification that a new paper has been submitted to their journal, and a new process will begin.

This process is peer review, but for many journals it will likely be preceded by an initial screening step to determine whether a paper is worthy of the time and effort that peer review will incur. Things that might be checked here by the editor with initial responsibility for handling the article include whether the paper falls within the scope of the journal; if authors send a paper outside the area of coverage of a journal, through not doing their homework carefully, they will likely get a quick email back saying, without a judgment on the scientific merit of their work, that they should submit it elsewhere on this basis. Papers that are very badly written may also be sent back with an invitation to rewrite and resubmit, if through the grime a glimmer of interest could be detected, but those which strike an editor as being simply bad science, whether well or poorly written, are likely to be rejected here. Editors may also seek a second opinion at this stage, from another editor or a member of their Editorial Board, to see whether the paper is worthy of review.

Papers will also, increasingly routinely, be checked for plagiarism at this stage. For many journals a standard process is for the submitted manuscript to be electronically compared to a huge database of published works to highlight the extent of any similarity to previously published work by other (or the same) authors. The problem of plagiarism will be discussed in more detail in Chapter 7, but for now we can say that the role of the editor is to check the outcome of this process and see if any issues that become apparent in this regard are of sufficient seriousness that the paper be rejected on this basis alone, or if areas of overlap are evident but relatively minor, and can be dealt with at a later stage if the paper clears the hurdles of peer review. The editor should be sensitive at this point to the result of the automated plagiarism analysis, as a simple percentage score doesn't capture the nature of the overlap (a few sentences here and there could, possibly, be coincidence, whereas a whole paragraph (or more) lifted from a single source is not, even if the Infinite Monkey Theorem applied.[2] A colleague of mine once had a paper rejected based on a journal's software detecting a huge level of text overlap in their paper to the thesis of the student first author which it found in our university's online thesis repository.[3]

In my own journal, around 50% of submitted manuscripts do not clear this initial prescreening hurdle and are rejected before peer review for one reason or another. Sometimes this saves authors time and sometimes it saves later embarrassment, and we will consider later the options authors have for responding to negative outcomes of all stages of review.

The Identification of Referees

In the past, there may have been a time when fields of science were such that editors of journals felt sufficiently expert in all aspects of research in that field that

---

[2] This concerns the odds that a room full of an infinite number of monkeys with typewriters could produce the complete works of Shakespeare if given an infinite amount of time in which to do it (and presumably an infinite supply of treats as well). As Han Solo famously said, though, "never tell me the odds."

[3] It is generally accepted that a thesis is an examination document and not a publication, and so in this case it would be entirely expected and uncontroversial that a paper extracted from the work described in a thesis would strongly resemble the parent work.

they could confidently judge in detail the scientific quality of every paper submitted to their journal.

However, as fields grew and complexity increased, the chances that any individual could judge the specific details, merits, and context of every paper they received, even in narrow and defined fields of research, became miniscule. So, today, every editor relies on the advice of experts in the specific area of a submitted paper before they make a decision as to whether the paper should be accepted, rejected, or modified.

This is the process of peer review, which in practice means that journals send a submitted paper to several experts (probably two on average, but maybe more) in the field to judge the paper's merits and decide on its worthiness for publication. These referees are in most cases anonymous to the paper's authors, who receive only their reports and the editor's decision, informed by these reports, as to whether the paper has been accepted, rejected, or is being sent back for further clarification and resubmission.

Peer review is a critical element of the world of scientific research, both in terms of quality control and the responsibilities of researchers. Essentially, peer review is a crowdsourced means of regulation of the quality of research published in a field of science, whereby the community itself regulates what gets published, and key recommendations are ideally made by those in the best position to judge the merits of a submitted paper.

So, when an editor receives a new paper, one of their first tasks is to consider who would be good, qualified reviewers for that paper. How might they find such individuals?

Part of the expectation of a good editor is that, through experience, they will accumulate a good feel for the field of their journal, and experience of reviewer (and author) behavior such that they can use multiple tools to supplement their own first instinct (if they have one) as to who might be suitable reviewers for a new paper. Typically, the tools they might use would include the following:

Internal journal databases: Over time, a journal will engage a huge number of researchers in roles of authors or reviewers, and the platform systems mentioned earlier will track all such engagements, with each individual also having indicated their areas of expertise (which they are usually requested to fill in by the journal at an early point of interaction), leading to accumulated intelligence on a huge number of potential reviewers. From this resource, an editor can typically search for all reviewers or authors who have previously indicated they have experience in key areas associated with a newly submitted manuscript, and also see whether they have reviewed previously, how often, when they did so most recently, and other such information, and this will often be a key step in identifying potential reviewers.

Literature searches: Editors will read the submitted papers sufficiently to be able to conduct their own searches on major databases, looking for authors who have published papers relevant to the topic of the submitted manuscript. Another way to do this is obviously to work from the papers the authors of the new paper have cited in their reference list as being relevant to their work.

Author suggestions: As mentioned earlier, most journals will request authors to suggest reviewers for their manuscript, but as editor myself I tend to be somewhat suspicious of these, unless they happen to correspond to whom I might have identified myself. Suspicious moves by authors include suggesting reviewers who are in their own institutions (a real rookie mistake) or even, if in small communities, within the same country.

Through some or all of these routes, editors should be able to construct a "short list" of potential reviewers and then start to issue invitations. This can happen quite readily using the online platform, which will send those individuals emails that usually include the title and abstract of the paper and a request to indicate their willingness to review by an indicative deadline.[4]

It is interesting to consider the basis on which this request is made; is it a professional request or a personal entreaty, a plea to a researcher's sense of duty to the community or a simple and easy to decline impersonal request not from a person but an electronic behemoth of a system? I believe that it is (or should be) more difficult for someone to decline a more personalized invitation than what appears to be an impersonal electronic invite; the guilt factor in saying no in writing should be greater than that required to simply "click this link to decline," and so as editor I take as much care as possible to personalize review invitations and make as clear as possible that there is a real person, with expectations and a need to do a professional job, behind the links.

This is all because the greatest frustration of most journal editors is probably the reviewer who ignores an invitation and wastes time for everyone as a consequence. Any editor would say that the best reviewers are those who accept requests quickly and then submit a good review on time, but I also acknowledge those who say no immediately and let me know that I need to move on to the next on my list. Those who ignore the original invitations, and then the reminders to consider that invitation, just lead to delays in the process for the authors, and are not any editor's favorite people.

Ways around this could include inviting more than the required number of reviewers, for example inviting four immediately on receipt of the paper in the hope that two will accept and do the necessary work. However, as editor I prefer not to do this, as it involves bombarding reviewers with more invites than needed, and I would rather hope journals could be in a position to respect the incessant demands on referees' time and trouble them as infrequently as possible, leading to a greater likelihood of an acceptance when asked.

---

[4] One journal has the peculiar (and unacceptable) practice of regularly sending me as email attachment the entire paper submitted, with the confusing accompanying request only to respond if I was willing to review the paper (so that, if I didn't, I still had the author's full paper without any further responsibility).

It is sometimes said that if you need a job done you should give it to a busy person, and I think this applies particularly to refereeing, as, the more busy and successful a researcher is, the more likely they will attract invitations to review, from a wide range of sources, some far more enticing and reputable than others. If such researchers are bombarded with invitations, they are less likely to have time to accept any individual one, no matter how well intentioned they may be, and so in principle journals should be as respectful of this as possible.

However, for that position to prevail, then referees should indicate their yes/no response to an invitation as quickly as possible.

Needless to say, from time to time researchers may be sent papers that are clearly outside their area of expertise, in which case they are doing no favors by accepting the invite and should just decline immediately; I have personally been invited to review papers on friction in aircraft tires, for example.

Sometimes, I have as editor had to send up to 10 invitations to get two reports, had papers of mine returned post-review by journals with apologies from the editor that they couldn't get more than one report despite multiple invitations, and had other papers where 4–5 reports on my paper were returned, sometimes with numbering showing the number of invitations sent (i.e., reports labeled as coming from reviewers 1, 3, 6, and 9). These all indicate the difficulty of journals in finding good reviewers, and emphasize that researchers need to be constantly mindful that, for the system to work for their papers when they need it to, they need to contribute themselves because, if everyone were too busy to review papers, two equally awful scenarios could result: (1) nothing gets published or (2) everything gets published.

In the month in which I wrote this chapter, I received nine invitations, three from the same journal, and in each case thought that would be the last for now, as I had enough on my plate, but each subsequent invitation indicated a paper that sufficiently tweaked my interest or related so closely to my own work that curiosity overruled my time constraints; of the nine, I accepted seven.

As a final key question to consider here, why do researchers review papers?

When I ask this question at workshops, participants often suggest that it allows an opportunity to see work in your field in advance of publication, but (1) obviously not everything that goes for review will be of high quality, and all reviewers spend time on bad papers that end up rejected and so do not present an early glimpse of anything useful, and (2) reviewers are bound by confidentiality and so cannot do anything useful with the knowledge they have been privileged to see before anyone else. Reviewing does give experience of applying the same critical faculties that will be applied to one's own work, which is useful, and also stating that one has reviewed for respected journals is an achievement sufficiently significant to warrant mention on researchers CVs and other materials relating to career progression, which is another benefit.

However, these are not the prime reasons why researchers should review papers. That relates to a basic matter of professional responsibility arising from being a member of a scientific community.

Every researcher is subject to peer review when they submit a paper to a journal, and should benefit from the experience, ideally because their work has been judged and hopefully deemed suitable for publication, as well as hopefully receiving affirmative positive comments on the work. Also, the feedback the reviewers have provided (detailed advice from experts in the field, provided for free) should improve the work or provide benefits for future work, even if the outcome in the specific paper concerned was rejection. Thus, every researcher feels the impact of, and benefits from, peer review (which is also applied to their grant applications, proposals for conference presentations, books, and many other forms of research activity), and it is a completely fair expectation that every researcher contribute to the process in return.

It would be extremely hypocritical for researchers to expect others to give their time, energy, and intellectual contributions to their work, but yet, when asked, refuse to contribute to the activity for others. There is never obviously a direct return of "you review my paper, and in return I review yours" (or at least if there is it could be coincidental and is not apparent thanks to the anonymous nature of the peer review process), but it doesn't seem unreasonable to say that, if two referees review every paper a researcher submits then, to "balance the books," researchers should review twice as many papers as they publish.

I have heard an argument that this responsibility should be modified by the number of authors on a paper, so that someone who was an author on a paper with ten authors only needed to "pay back" one-tenth of that in reviewing duties; I do not agree with this as, first, every author gets the benefit equally for their CV and career and, second, many authors may not be in a position to contribute in the future to peer review (for example, if they are students who go on to nonresearch careers, or support staff whose roles do not allow time for such duties).

Nonetheless, I think it is fair to say that, for most editors, one of the most frustrating parts of their job is still finding referees.

Interestingly, a number of initiatives and sites have arisen that seek to recognize the efforts of reviewers more explicitly, which may help offer additional incentives for researchers to participate in this critical (literal) activity. These include the Peerage of Science, Reviewer Credits, and Publons; these seek to turn peer reviews into measurable research outputs, in some cases "review the reviews," and provide training and support for reviewers.

The Hacking of Peer Review

There have been a few reports in recent years of authors attempting to "hack" the peer review process in their favor. Early examples of this involved their suggesting researchers who did not exist as referees, with the email address provided being controlled by the author, who then did a review of their own work (and presumably found it excellent!). When cases like this came to light, the editors responsible, who had accepted reviewer suggestions without even checking that they were real people, lost their positions for negligence.

In a later version of this scam, authors proposed the names of real researchers who would likely be known to editors, but gave for them email addresses that were (you guessed it) fake and controlled by the authors, who then were able to again favorably review their own work.

It is unfortunate that editors need today to be attuned to such possible malfeasance by authors ("fake" reviews) but, as with many other areas of publication ethics, a few cases of extraordinarily inappropriate behavior by a few have resulted in greater scrutiny and suspicion for the vast bulk (we hope) of honest authors.

More ambitious attempts to hack the review system include reports of breaches of security in the online editorial systems used by publishers, with attempts to steal passwords and log-in details for editors and reviewers, so as to manipulate the review process electronically.

The Roles of Reviewers

What do editors expect and request of reviewers who do accept an invitation to review a paper?

A simple phrase being used increasingly, which is a good place to start, is "sound science." Is the paper a useful, original, well-conducted contribution to the field of research, which reports findings that others will find useful or interesting?

Editors should not expect referees to do detailed editorial checks of papers and fix grammar and typos; ideally, it should be the responsibility of authors to submit papers that do not need such effort to be expended, and of editors to enforce this by not making a referee's life harder through sending them a paper that they will struggle to read.

Small things like document formatting, grammar, and lack of typos are a huge part of making a professional impression on the editor and reviewers. A reviewer seeing an annoyingly badly presented paper will immediately get a bad impression of the authors through simple human instinct even before they engage with their science, and so taking care in this respect is a simple and obvious way for authors to make a positive first impression.

I once heard a very experienced recruitment professional say that many job interviews are won or lost in the first minute, and I know that interview boards make immediate initial decisions on a candidate before they open their mouth, based on how they have dressed, composed themselves, walked in, stood, made eye contact, and gave a million other nonverbal clues as to their preparation, personality, and suitability. A sloppily prepared manuscript is like appearing at interview in dirty clothes, smelling of booze, having slept the night before in a ditch—it makes a bad impression and leaves the paper with several negative marks against it even before its contents are considered.

As author, reviewer, and editor, I have at this point seen probably over a thousand referee reports, and needless to say no two are the same. Some are short, some are long; some are superficial, some are incredibly deep and detailed. Most are helpful in some respects, many are polite and professional, but some are inappropriately personal and insulting, and some are simply wrong.

A recent paper of mine stimulated from a referee the following report, in full: "Be careful of page layout and edges in your article."

What an author needs from a review report is clear, fair, and unambiguous feedback on their work. If there are problems of such a level that rejection is recommended, then the rationale for this needs to be spelled out, so that the authors understand why such a harsh outcome has arisen. If changes need to be made, or questions answered, then the nature of these needs to be clear such that the required actions that will satisfy the reviewer and editor are clear.

What does an editor expect of a reviewer? The key point is that they expect that the reviewer has expertise and perspective that they will apply to the paper in question, and so consider the science in detail. Typical questions a reviewer should consider might be as follows:

- What is the point of the study, and was there a good reason for doing it at all? This captures points such as originality (is this really new?) and interest (will anyone care about the outcomes of the study?).
- If the study had a good and worthwhile rationale behind it, was it competently conducted? Were the methods (and materials) used appropriate, current, validated, and likely to lead to good data for the study conducted? If required, is sufficient evidence provided that the work was conducted in an ethical manner?
- Was the study designed and conducted in an appropriate way, for example in terms of replication, numbers of samples or subjects, and statistical analysis (if appropriate)?
- Do the authors understand the context of their work, as reflected in the papers they cite and how they relate their work to that of others, in particular in the Introduction and Discussion sections of their paper?
- Are the conclusions reached supported by the data?
- Was the paper written such that readers can understand and engage with its contents properly?
- Are figures and tables, where used, clearly laid out, useful, and necessary?

These questions should get to the heart of the case for publishing a paper, and provide the editor with sufficient expert advice to make a judgment on this point, and then provide the authors with a clear roadmap for progress, or else clear reasons for not progressing. Criticism, where possible, should be constructive, giving advice as to how to resolve problems identified.

To consider how a review might work, it is worth considering what a typical peer review assignment might look like (based on my own practice, which I am sure is not too atypical). When I have accepted a paper, I will have read the title and abstract, and so already have an idea of what the paper is about and whether I think it is interesting (hopefully so if I have accepted an invite to review it) and whether it is going to be a slog or not to read it (how the abstract is written alone will be enough to give a sense of the mud that needs to be waded through in search of treasure). First impressions will count!

I always print out the paper, as I will review it first with (red) pen in hand, typically check the deadline for review (usually 3 weeks in my field, but less in some others), and put the paper on my "to do" pile, which I check frequently for items with increasing levels of urgency. I usually take out the paper at least a week before the deadline (or more if I have a long flight or train journey on which I can catch up on such assignments).

When actually reading it, I often go first to the figures and tables to get a sense of what was done and found quickly, and then work through the paper page by page. Eventually, after probably two full read-throughs, annotating by pen each time, I will type up my report and then upload this to the journal review site (almost always on a day after the readings, so coming to it relatively fresh), making sure to present my global comments and impressions first, followed by point-by-point specifics. There will typically be a box in the review submission form for comments to the editor, which are not seen by the authors, but I try and minimize this, unless there are points that relate to the broader process to follow; for example, if I am somewhat "soft" in my recommendation, I will usually indicate that, if the other reviewer(s) and editor disagree with my judgment, I will not object and accept the consensus.

I pride myself on not being late with my reviews, but I estimate that at least a third of the reviews I receive as editor are late!

The Editor's Decision

Eventually, when the editor has found two (or more reviewers) to do the above and they have submitted their reports,[5] it is up to that editor to exercise their next key function (after finding the referees) and decide what to tell the authors.

The editor of a modern scientific journal manages a rather complex process; procedures that are probably typical of many journals are schematically illustrated in Figure 5.1. In the upper flow-chart, the processes that occur on receipt of a paper are shown. These processes can lead in a high proportion of cases to rejection, for a range of reasons, while the remaining papers are passed on to a handling editor, who then communicates with the reviewers (and the subset of these that form the Editorial Advisory Board, on which more follows), as summarized in the lower panel.

---

[5] Sometimes after a lot of chasing for them to do so, another frustrating part of an editor's job.

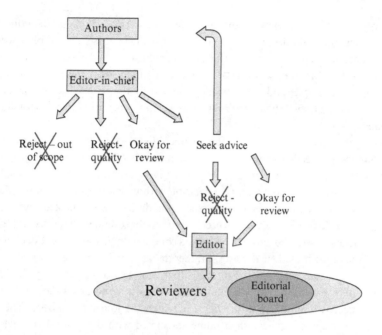

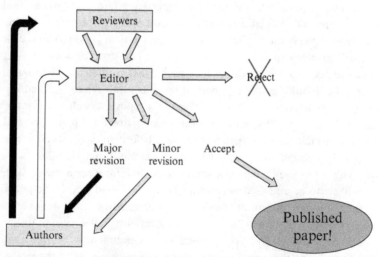

FIG. 5.1 *Flow charts representing the processing and peer review of papers in a typical scientific journal, showing prescreening (top) and full peer review (bottom). In the bottom diagram, the flows for major and minor revision are shown in black and white, respectively.*

The purpose of the review reports is to provide expert advice to the editor, on which basis they will decide the fate of the paper. Of course, the simplest scenario is where the referees come to the same conclusions, and those match that of the editor, for then the path forward is clear. If all consider the paper should be rejected, for example, and the reasons presented by the referees seem fair and valid and consistent with the editor's own impressions, then only one outcome is reasonable.

Things get more complicated if the referees disagree, with one saying only minor changes are needed while the other recommends rejection, for example, or where the reports are unclear or seemingly excessively superficial and lacking sufficient grounds for the editor to make a decision with which they are comfortable.

Where I, in my role as editor, encounter such quandaries, I will almost always seek additional input. Even if the paper has been in process for several weeks, it seems fairest to the authors to seek the most complete and coherent set of views before deciding the fate of the product of their hard work rather than acting in uncertain haste.

This can be where a journal's Editorial Board, or Editorial Advisory Board, comes into play. Every journal will publish in their issues and/or on their website a list of respected researchers in the fields in which the journal publishes. These individuals have agreed to have their name associated with the journal and get a modest benefit for their CV in terms of a mark of esteem among their scientific community, in that they are sufficiently respected to be asked to be so recognized; in return, they have implicitly or explicitly agreed to referee regularly for said journal, to be someone who editors know (if not excessively pestered) will not turn down a reasonable request to review a paper, and to do a good job on such assignments. It is to such experts that an editor might first turn for advice on the basis of sticky review reports, or to act as a tie-breaker in the case of differences of opinion. The Board member will be sent the paper and typically asked for a fast turnaround; sometimes, an editor might send the reports as well, to show the issues raised, or else they might suggest the Board member read the paper first and come to their own conclusions but review the reports before uploading their final results.

Hopefully, whatever the precise flow-chart steps followed, the editor will have a verdict with which they are happy, and be ready to inform the authors. If the outcome is likely to be a revised manuscript being submitted (i.e., if the judgment is other than rejection) the editor should at this point address those issues which they have advised the reviewers to ignore, such as formatting, adherence to the journal's instructions to authors, and writing issues such as grammar, typos, punctuation, et cetera. The editor will normally flag such matters at this point in the report going to authors; in my own case that involves printing out each paper, indicating all editorial changes and queries by (red) pen, and then scanning and uploading the resulting file for the authors to work from when preparing their revision.[6]

---

[6] I have occasionally received emails from puzzled authors all around the world seeking clarification on some of my scraggly hand-written notes, but surprisingly fewer than might be feared. I also accept that this is not common practice for editors.

Peer Review, Confidentiality and Collaboration

In the classical review methodology practiced by most journals, the editor knows who everyone involved is, the reviewers know who the authors are (as they see the paper), but the authors do not know who the reviewers are, and each reviewer does not know who the other reviewer(s) is/are. Why are there such variable practices for confidentiality and anonymity?

First, let's consider the position of the referees. What the editors expect from them, above all else, is a frank appraisal of any papers they are sent, for better or for worse. However, many fields or specific areas of research are small, and hence there is perceived (probably realistically) to be a risk that one researcher might be less inclined to say that a paper by someone they know through interactions within the field, even indirectly, is rubbish. Editors expect that, if someone gets a paper to review from their best friend and it is terrible, they will say so, and to help them do this without fear of possible future negative outcomes (and we could picture many forms of these) they are provided with the cloak of anonymity.[7]

Does this (so-called blind reviewing) have dangers? Possibly, in that the reviewers are coming into contact with privileged information before publication, without the author's knowledge, allowing for the possibility of theft or ideas or materials, or the application of rivalries or grudges in terms of rejecting the paper of a competitor, for example. For this reason, some journals operate open reviewing, where reviewers must sign their name openly to their reviews.

Many journals also have two open boxes in the online forms that reviewers must complete, one for comments to the authors (which they see) and one for comments to the editor, where any additional comment or perspective may be shared that could be helpful in reaching a decision (in my experience, this can often be ignored if the section for authors, as it should, stands by itself and is clear and complete, but I mentioned some exceptions earlier). However, some journals have also removed this "confidential note to the editor" option, to prevent referees from making positive comments in the author-facing section but undercutting or contradicting these by much more harsh comments to the editor; there is a good case to be made that it is fairer to the authors that all views of the referees be made known to them.

Are there any dangers to the reviewers knowing who the authors of a paper they have been sent to review are? Yes, because, as highlighted earlier, it has been suggested that sometimes reviewers may be inclined to be less critical or harsh on papers from researchers or institutions that are well respected or renowned (and indeed this has been suggested to be a factor in cases of fraudulent papers from

---

[7] I once had a paper the title of which changed during peer review (as can sometimes happen) and around the same time saw it referenced on a poster at a conference as (in press) with the original title. This was an unusually obvious indication of the role of the senior author on the poster in the review of my paper!

some of the top institutions in the world), while the reverse could apply to papers from less illustrious origins (Ireland? Never heard of it!).

For this reason, some journals operate blind peer review, where the front page of a manuscript with author details is not sent to peer reviewers. The principle here is that the authors bring no baggage with them into the review process, in terms of who they are, where they come from (so that publishing moves closer to a real meritocracy, where what you have to say is all that matters, rather than who you are), or even their gender, and there is probably a very good argument that this should be more consistent practice.

Finally, what about the fact that reviewers work in isolation, without knowing who the other reviewers are or were for a paper they are working on? A possible way to avoid reviewer disagreements is for the reviewers to confer with each other, in what is called collaborative peer-review. In such practice, reviewers are in contact or at least aware of each other's comments and views (perhaps anonymously), and can discuss their views and reach a consensus, such that authors should get more consistent and less potentially contradictory feedback.

While this has been experimented with by some publishers and journals, I think there is richness in different independent views on a paper, and the role of the editor is to achieve consistency and clarity for authors in terms of what is expected of them. Reviewers may disagree, but in such cases further reviewers can be sought (albeit at the risk of delaying the process), and there is also a risk in collaborative review that one referee can change the mind or influence the views of others, and that a sort of "groupthink" could emerge; in extreme cases, an editor could find themselves trying to mediate an argument between disagreeing referees!

In some journals, a mixed model is used, where there is a first independent review phase followed, when the independent review reports are available, by an interactive review phase, where a forum allows direct contact between authors, reviewers, and editors.

The broad term "open peer review" is today being used to refer to reviewing models where transparency of the review process is increased by practices such as having nonblinded (and possibly collaborative) reviewing or publication of review reports alongside published papers (some journals offer this as an option for authors to voluntarily agree to), or both. The concept of "openness" may be even broader, with readers being allowed to contribute to the critical evaluation of manuscripts post-publication, and there has also been discussion of manuscripts being made available in advance of peer review (which seems strange, as a lot of such manuscripts will likely be rejected for flaws that render their having been made available earlier of dubious value). Multiple sites offering such services have been established, including Academic Karma, Haldane's Sieve (for discussion of genetics papers), and Peer Community In (PCI), which seeks to offer a community-based platform for review and recommendation of scientific papers. The British Medical Journal allows nonanonymous prepublication peer review, alongside a multitude of different peer review models (from fully blind to fully open).

So, essentially every combination of interaction, from fully open systems to fully blind systems where everyone or no one knows who the other players are have been tested and are in practice somewhere, but the semiopen model (authors known, reviewers anonymous) remains the most common.

Responding to the Reviews

Researchers of a certain age (hey, not that old!) will still remember the sensation of receiving a posted report on one of their papers from a journal, and the nerve-wracking experience of tearing open the journal-logo-bearing envelope to learn the outcome, like a personalized Oscar ceremony. It just doesn't feel the same when the report comes by email!

Whatever the moment of delivery, the first text encountered should be the response of the editor, and will be quickly scanned for critical words like "Major," "Minor" or (gulp) "Reject." This will be followed by the text of the referee reports, which should be consistent with the headline judgment and provide the details of the overall verdict of these anonymous experts and their recommendations, including points for amendment or clarification should the verdict not be an outright rejection.

Unless the verdict is very benign, a common approach I take is to set aside the reports for a little while, from an hour to an overnight, to let immediate responses of anger, confusion, or disappointment subside and allow for a more dispassionate evaluation of the reports and required next steps to be undertaken (presumably after circulating same to all coauthors to solicit their input).[8]

What actions are possible or options are typically open? This depends on the headline judgment, as follows:

Accept as is: In all my experience, this will never happen for an initial submission of an article as, even if one referee believes the paper to be perfect as it stands, this is highly unlikely to be shared by the other referee(s) and editor. The goal of course is to iteratively approach this result through revision and cycles of re-review and engagement with the editor, but as a first-round result it can be effectively discounted.

Accept after minor or major revision: Outcomes where the authors are offered the opportunity to revise and resubmit their paper will either fall into the category of minor or major revision. The difference here, as most journals will approach it,

---

[8] A very nice example of a researcher's response to a negative review was posted on Twitter by a behavioral ecologist called Kirsty McLeod on September 19, 2018, and spurred a good thread of researchers' accounts of similar experiences. The thread can be seen at https://twitter.com/kirstyjean/status/1042458943870709761.

is not the exact number of queries or their severity, or requirement for additional work to be undertaken, but the process implications. If a reviewer recommends minor revision they are saying that, if the recommended changes are made, the editor can review the new draft and decide on the acceptability of the paper without the need for further review.

In contrast, a recommendation for major revisions usually means that the reviewer needs clarifications and improvements to be made before they can decide whether the paper is publishable or not. In some cases, the editor may decide whether the reviewer's concerns have been adequately met, if they feel sufficiently expert in the topic to make this judgment, without further review. Generally, though, the revised and resubmitted paper will be sent to the same (and possibly new) referees for their advice, and the paper may be rejected at that point ("now I understand better what the authors have done, it is worse than I originally thought . . .") or even have to go through further cycles or revision and review (as editor, I have had papers go through four or more cycles of peer review before acceptance).

In either case, when preparing a revised paper, the authors need to prepare a separate and new document that makes it very clear what steps they have taken in response to the review reports. This might, for example, take the form of a table or list of reviewer remarks, opposite or below which the authors answer the questions or criticisms raised in the review, and indicate what recommended changes have been made and, if any were not made, provide a rebuttal as to why not.

Of course, even before undertaking such a detailed response, an immediate question is whether the authors wish (if offered the opportunity) to revise and resubmit their paper. For example, if the referees recommend that the paper following revision include new data or if addressing their queries or criticisms would require further lab work, this presents the authors with a key decision as to whether such work is possible or desirable (for example, the student(s) involved may have finished up, budgets may be used up, or methods or equipment required might not be accessible). In such a case, there should be no problem with the authors explaining the situation to the editor and requesting to withdraw their paper from consideration on this basis.

Rejection: As stated several times in this book already, most journals of high standing have a very high rejection rate for papers they receive, and every researcher has had the experience of having their papers rejected at some point (even if they try and deny this). In every case, this is a dispiriting and disappointing experience, and days on which such letters or emails are received are never the highpoint of anyone's week.

Some comfort may be taken from the fact that there are several examples of famous, even Nobel Prize–winning, discoveries that were at first rejected by the journal to which they were submitted, including the first descriptions of the Krebs cycle of metabolism and the polymerase chain reaction (PCR), a commonly used tool in molecular biology, and Peter Higgs's first paper on what came to be called the Higgs Boson.

When a researcher has a paper rejected, having calmed down and taken a breather, as advised previously, the next question in such cases is what to do next. Typically, there are three obvious courses of action open to the authors:

1. Submit the paper to another journal. As stated earlier, a journal will have been selected for submission from a list of possible suitable ones, and while parallel submission to several journals is not acceptable practice, sequential submission to another journal on the list certainly is, typically working downward in criteria such as Impact Factor (no one will get a paper rejected by a modest journal and think "to hell with it, let's send it to Nature"!). Before doing so, however, the authors need to be careful to do two things.

   First, they need to ensure the paper is reformatted for the next journal chosen, as nothing is as easily detected and as frustrating to editors as a paper that clearly bears the formatting fingerprints of another rival journal.

   The second thing is to take advantage of the expert consultancy the authors have received, for free, from the reviewers of the first journal, and fix as many issues as can be easily addressed (assuming that there are some killer criticisms raised which, as the basis of rejection, cannot be fixed). This type of revision between submissions makes perfect sense, not only because there is probably a nonzero chance that the second journal will send the paper to one or more of the original reviewers, who might look more favorably if they see an effort has been made to address their criticisms.

2. Do not submit the paper to another journal. This is obviously the hardest decision to make, but if the referees raise fundamental flaws with a study that the authors had not foreseen, there is probably a high chance other reviewers in a different journal will have the same problems with the paper and that further rejection is a likely later second blow. In such cases, it may be best to simply accept that publication is never likely to result, and save everyone work, time, and emotional trauma by not proceeding further. This experience should at the very least be a learning experience, from which lessons for future studies can be extracted. On at least one occasion, I have essentially repeated a study based on referees' feedback on a previous version, changing methods and design, and had the later study published in the original journal (having been careful to point out the relationship between the studies in the covering letter to the editor).

3. Negotiate with the rejecting journal: This is a path I have chosen a small number of times in my career, and in each case approached with the greatest care and diplomacy, but I think all eventually resulted in publication of the original paper in that journal. The key grounds for initiating such a dialogue is obviously that the referees misunderstood a key aspect (or aspects) of the original paper where that aspect was the basis for

rejection, and where a reasoned argument could explain why that was not a fair basis for the verdict ultimately reached.

For example, in one case a reviewer was quite nasty about our paper, and said that it was not original, referencing as support for this assertion the work of a particular researcher Dr X. As soon as I read those words, I knew immediately (1) who the referee was (not Dr X) and (2) that they had completely misread the paper, as the work of Dr X was on a topic related to, but completely different from, that in our paper. Once a calm and reasoned (I cannot stress those words enough!) approach was made to the editor to politely point this out (the wording used was, for sure, quite different from that uttered on reading the report), the paper was published.

In another case, three referees reviewed one of our papers, and two recommended minor revision while one rejected the paper, and the editor sided with the rejector, although their basis for rejection was a rather vague comment about the work representing a minor advance from what was known. My coauthors and I spent a lot of time carefully crafting an argument as to why we believed that was not a fair or accurate interpretation, and showing how we had progressed beyond previous work quite significantly. The editor then responded to say the paper remained rejected, but that based on our arguments they could see a basis for a new submission in which the advance was foregrounded more clearly; through the crack thus opened in the door, we pressed our revised paper, and publication ensued.

To choose this path, an author(s) must have a very clear argument to make, and then make it in as dispassionate and professional a manner as possible. An email (or worse phone) rant at an editor about a favored paper being rejected is unlikely to win them to your side (which is essentially the goal here).

In my career, I would estimate that I have had, out of around 250 research papers on which I was author or coauthor, no papers accepted as submitted (no one has!), around 10%–15% rejected (and all bar a handful of which were published later either in the same journal or another), and the rest split around 60/40 between minor and major revisions. I am not sure how that matches typical experience, but is probably somewhat better on average than stats for a journal (where probably far more than 50% of submitted papers get rejected).

What is the best way to avoid rejection? Besides the obvious (do good science, based on a good idea, come up with an interesting and original conclusion, and then write good papers about the outcomes), I think key advice is around care in preparation before submission, with as many drafts as required (whether 3 or 13) being taken to make sure the referees and editor are presented with a manuscript that is professionally and carefully polished, and that will make their task of judging the substance behind that style as painless as possible.

The outcomes that can arise from peer review, and the workflows and steps that can follow, are summarized in the lower part of Figure 5.1.

Acceptance and Publication

While it may seem like a very high proportion of papers submitted to journals get rejected (and indeed they do) of course a proportion do get accepted, hopefully the best proportion in terms of significance and reliability. These papers may be accepted in a form very different from that in which they have been submitted, having been molded through the unsung intellectual collaboration that peer review can provide, and may even be accepted in a journal different from that to which they were first submitted.

Whichever the route, the authors of such papers will be notified that their paper has been accepted, and a new and separate process will begin, with which the editors generally have little direct involvement.

One key process that occurs at this stage regards the formal steps around copyright assignment and publication charges, whereby authors either (in the traditional model) sign off their copyright for the paper to the publishers (and in some of these cases still indicate how they will settle the accompanying publication charges) or agree to retain same, for a fee, in an open-access publication agreement.

The next main step is the preparation and checking of proofs of the paper. While some journals now require authors to submit their papers in a format resembling that in which it will finally appear in print, in most cases what has been accepted is a word-processed file with separate figures and tables, which requires assembly into a format that looks like a published paper. For most publishers, this is done by professional copyeditors and typesetters, who take the authors' files and lay them out in what is called a *proof*, or mock-up of the final published article.[9] These are then sent to authors for checking, to confirm that the proposed version to appear is a true reflection of what they intend to publish.

Proofs are traditionally sent with very short timelines for checking and return, such as 48 hours, and it occasionally feels like publishers watch authors carefully to check the absolute worst time at which to send them. I once received five sets of proofs together (for a special issue of a journal arising from a conference I had organized) at noon on Christmas Eve, with a 48-hour deadline for return (clearly, at least I hope, an outcome of an automated process, rather than a publishing Grinch)!

A critical point is that no scientific changes are allowed at print stage, as the editorial process for such matters is complete, and so changes like fixing a claimed typo that alters "has an effect" to "has no effect" are forbidden, as are changes to author lists lest issues relating to research integrity arise at this point. Nonetheless, in my experience, no proof does not require at least some minor changes, from fixing typos or updating details on references to, at the larger end of problems, noting that the figures that appear in the proof are not the ones

---

[9] These used to be called *galley proofs*, a centuries-old reference to the original design and operation of printing presses, in which type was laid in trays called galleys.

supplied for that paper or that one figure appears twice (both of which have happened to me).

There used to be an arcane system of codes and marks used for marking proofs by hand with pen or pencil, with particular symbols to be used that indicated where insertions, deletions, or rearrangements took place. I learned these when I started publishing, and still use them in my hand-editing of papers, but today these have been largely superseded by electronic annotation of PDF proofs.

To speed up publication, or at least the access to information, most journals now make proofs available online in advance of the final prepared version being available, which means that a paper can be available within days of it being accepted for publication. However, even this is changing today, with journals (those that still print hard copies) assigning issue and page numbers quickly on receipt of proof corrections so that the final version is perhaps available long ahead of its printed counterpart, while, for those journals that exist online only, the online copy that appears quickly is of course the final version.

When I started publishing papers a further step was the ordering of what were called *off-prints* of the paper, and authors generally got 50 free hard copies of their paper on nice shiny journal-y paper to distribute to anyone who may be interested, but of course the emergence of paper sharing electronically by PDF has relegated such practice to mere historical curiosity (plus many large piles of such off-prints for older papers in my office!).

So, we have seen how a paper goes from submission through the process of peer review, to a range of possible fates, from rejection to eventual acceptance, and appearance for the first time in the world as a new contribution to knowledge. What happens next?

6

What Happens after Publication?

TRACKING THE IMPACT OF PAPERS

A key term used in recent years in relation to the ultimate outcome of research is "impact," which is defined by the *Oxford English Dictionary* as "a marked effect or influence" and was once defined in the UK as "an effect on, change or benefit to the economy, society, culture, public policy or services, health, the environment or quality of life, beyond academia."

Once a paper has been accepted for publication in a journal and passed on to the publisher, and a proof has been prepared and approved by the authors, the paper, for the first time, appears "in public," and can have an impact. For this contribution to be genuinely useful, it must impact on the work of others in the field, its findings being picked up by others and incorporated into and influencing their work.

In other words, a paper must be a link in the chain of knowledge, and thus must not stand alone, but be integrated into the bigger framework of research, and provide a step forward that is acknowledged as such by those who build on it.

In some cases, the world can get a preview of the work even before publication, as articles may appear on "preprint servers," websites where manuscripts appear online in advance of formal publication, and in some cases even before peer review; an example of a massive online site for preprints, initially focused on high-energy physics but now much broader, is called arXiv.org.

The Significance of Citation

For decades, one of the major records of a scientist's career and productivity was the list of their publications, and the standing of the journals in which those papers appeared (for example, it would sometimes be quipped that a paper in *Nature* was equal to 10 or more in lesser journals). Today, however, an additional measure of quality is required by those assessing a scientist's achievements (say in evaluation for a job or promotion): the number of citations to the work.

A citation is counted every time a paper is included by a later paper in its reference list, which will probably happen either in their summary of the state of the art in the Introduction, in supporting the use of a particular material or technique in Materials and Methods, in explaining new results, or in showing how understanding has moved forward in the Discussion. For whichever reason, this cross-referencing by the later paper constitutes a citation, and, when the citing paper appears, various databases will track the reference and increase the citation count for the earlier paper by one. With the advent of electronic literature databases, it has become really easy to evaluate the impact of papers in the years following publication, by generating in seconds lists of papers that subsequently referenced the earlier work.

More and more, the impact of a paper is evaluated by counting the number of other papers that subsequently cite it. A single paper with hundreds of citations will weigh more heavily than dozens that have no or few citations, as clearly this indicates that it was read and considered far more widely.

It seems reasonable to assume that citations correlate with impact of a paper, as a paper that drew no attention will attract no later citations. Of course, a paper may also refer to an earlier paper negatively, perhaps to say that the findings of the earlier work were wrong, as science exerts its own process of self-correction, and so not every case of a citation represents a tribute to the earlier work.[1]

Of course, there are reasons for low numbers of citations which cannot be ignored. For example, the more people are working in a specific field, the more likely that a useful paper will receive multiple citations. Conversely, in smaller disciplines, even well-recognized papers simply do not have the audience size to accumulate huge numbers of citations.

It should be noted that apparent attempts to manipulate one's citation scores, such as exaggerated self-citation (where an author references their own work when that of others could have been equally appropriate), are recognized and disapproved of. This practice can happen also at the level of journals, with there having been a number of cases of editors being reprimanded or forced to resign (even to the level of journals being temporarily withdrawn from lists of Impact Factors) for encouraging or demanding of authors of accepted papers to increase the number of citations to papers in that journal, blatantly and transparently trying to boost their Impact Factor.

---

[1] In one amusing case, a paper was published in the journal *Ethology* in 2014 titled "Variation in Melanism and Female Preference in Proximate but Ecologically Distinct Environments." Between authors, reviewers, an editor, and the journal publishing team, it would be expected that many pairs of eyes scrutinized this paper, yet none of these noted a comment in the Introduction which said, "could be a consequence of either mating or shoaling preferences in the different female groups investigated (should we cite the crappy Gabor paper here)." This note, which captures evocatively the internal discussions and perhaps rivalries that might exist among any group of authors while drafting a paper, was apparently added after peer review, and resulted in the withdrawal of the paper ("due to the inclusion of an author's note not intended for publication"). It was replaced with a new version in which that note has been deleted and one reference has been added (to a paper by Gabor in 1999).

Papers that describe new and useful methods are classically citation-gobblers; each time the method is used in a subsequent paper, the original paper is cited, and citations pile up just as royalties for Beatles songs mount up with each play on the radio.

Review papers, where the findings of dozens or hundreds of scientific papers are meta-analyzed, are also good citation-winners, as they prove a useful short-cut for referencing future literature reviews.

Papers do not generally go on accumulating citations forever; citations usually reach a peak and decline thereafter. This is because new findings, if important, eventually become so accepted that, if subsequent research has supported and built on the findings, they become such a part of common knowledge that the original source need no longer be cited.

This leads to an important distinction between *statements* and *facts* in science. A statement is something about which possible uncertainty remains; statements are open to disagreement, disproof, or argument, and typically need the moral support of relevant citations to the source of such evolving knowledge. A fact, on the other hand, is simply a fact, and does not need to rely on the credibility of those who first uttered it to establish its veracity. There are obviously no such things as "alternative facts" in science.

The way in which subsequent papers refer to a particular proposition can track the evolution of that proposition from statement to fact, or not.

For example, we could envisage the following series of statements appearing in chronologically consecutive papers (not actual quotes, but created for illustrative purposes):

It has recently been proposed (Watson and Crick, 1952) that DNA is a double helix.

These results support the double-helix model of Watson and Crick (1952).

DNA has a double helix structure (Watson and Crick, 1952).

The structure of DNA is a double helix.

On the other hand, we could envisage a different trajectory for a different discovery, to be explored later in this chapter:

It has been proposed that analysis of the Antarctic meteorite AL84001 indicates evidence of microscopic forms of life (McKay et al., 1996).

These results do not support the findings of McKay et al. (1996) regarding . . . .

It is no longer accepted that analysis of Antarctic meteorite AL84001 showed evidence of early life on Mars.

Of course, it is not the case that what are at one time accepted facts will not be later found to actually be wrong (for example, it was—and in some places still is—believed that the earth is flat), and many major discoveries in science have come about precisely because some doubted the veracity of specific facts, and were not afraid to tear down a particular edifice of belief (even if it meant taking on the conventional orthodoxy, or incurring the wrath of the Spanish Inquisition).

These issues lead us uneasily into the field of the philosophy of science, around the perimeter fence of which I will only tiptoe cautiously here. This is the discipline that seeks to understand the methods by which scientists operate and establish facts. While the legendary physicist Richard Feynman said that the philosophy of science was as useful to scientists as ornithology is to birds, many interesting ideas and proposals have emerged from this highly specialized field, particularly during the 20th century. Regarding the idea that scientific proposals (or hypotheses) must be capable of being proven wrong, the Austrian-born philosopher Karl Popper said that theories in the natural sciences can never be proven, but can only be disproven, and should be scrutinized by decisive experiments. Popper also said that a theory is genuinely scientific only if it is possible in principle to establish that it is false (an argument used sometimes to demarcate the difference between science and religion).

Popper's arch-rival, Thomas Kuhn, described revolutionary changes of mind within entire fields of science as "paradigm shifts," as exemplified by the change from a model of the universe that revolved around the earth to one in which the planets of the Solar System revolve around the Sun. These sudden shifts in understanding are followed by periods of "normal science," when researchers enlarge and fill in details of the paradigm, such as when a new discovery that changes the understanding of an aspect of science causes waves of subsequent work, debate, and perhaps confirmation or disproof by others. We will consider some examples of how discoveries permeate from the original publication in this chapter.

The citations to any paper can be regarded as ripples in the field in which they are published. Figure 6.1 schematically illustrates how the paper at the center (A) is cited by other papers (labeled B), each of which is cited by others, and these by others still (rings C and D). Perhaps by the stage of secondary and tertiary citations (C and D) the original paper is no longer being cited, but its influence is still being clearly felt.

It is easy using databases to track citation numbers over time, and several such trends will be discussed in this chapter. In addition, software exists today (such as one called VOSviewer) that can plot citation trees and maps, showing linkages between citing papers and groups. An example of such a visualization is shown in Figure 6.2, for a review article in my own field (in this case, the biochemistry of all the reactions that take place during cheese ripening that lead to development of flavors and texture characteristic of cheese varieties). The upper plot shows the links to papers that cited the original review after publication, clustered by shading into papers that interlinked in their citations.

In the middle plot is shown the coauthorship patterns of the author of that review, showing different strands of collaboration and joint publication, most of which are geographically clustered (like an Italian cluster at top middle, and a US cluster of papers at bottom middle). In these plots, the size of the circles represents

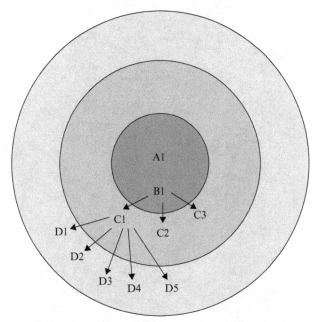

FIG. 6.1 *The ripples of citation from a single paper, as the initial paper (A1) accumulates primary citations (B1 . . .), each of which accumulated secondary (C) and tertiary (D) citations, so that the rings of influence of the paper grow progressively larger.*

the number of publications for named authors, while the authors that are most closely connected to each other (through such citation networks) have the same shading for their circles.

The same analysis based on key words linked to citations (the bottom plot) shows a huge range of terms, but these are clustered into key themes as different areas covered in the review were picked up in later papers relating to those areas, which led to other citations to those papers, and so forth.

There have been a number of controversial studies in the 1990s of megapatterns of paper citation trends that suggested that up to half of all papers published are never cited, but a 2018 article in *Nature*[2] concluded that the true figure for this may be as low as 10% (and gradually decreasing), and noted that citations are not necessarily correlated with downloads (and thus presumably the number of times a paper is read). They gave an example of a paper that was viewed more than 1,500 times, downloaded more than 500 times, and yet never formally cited. In addition, papers can clearly be read and used by others (and hence have unarguable value) without this being reflected by citations (for example, a paper that leads to a development in industry or public policy). However, while it is possible and perhaps inevitable that occasional low rates of citation might happen for papers presenting

---

[2] Van Noorden, R. The science that's never been cited. *Nature*, December 13, 2018

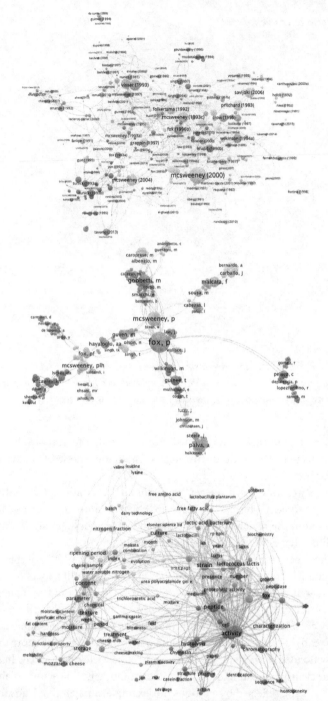

FIG. 6.2 *Citation maps for a review article on cheese ripening (Fox, 1989). (Top) map of papers that cited this original paper, (middle), map of coauthors of the main author, in which strands of collaboration and cross-citation (like conversations spinning off at a party) can be tracked going in several different defined threads of citation, and (bottom) a map drawn based on terms used in citing papers, in which key themes for which it was cited can be extracted, including proteins, microbiology, quality, and analytical terms. Maps were generated using the VOSviewer software.*

good science in what could be termed "proper journals," it is hard to imagine that papers being published in extremely dubious minor journals that have sprung up like weeds in the garden of scientific publishing in recent years will achieve a "most are cited" pattern.

A further aspect of writing in science is that science moves so rapidly that papers have a period of currency or relevance that can be quite short, before their content, if sufficiently important, becomes part of the day-to-day knowledge of people in the relevant field (and statements become accepted facts). The chemist Carl Djerassi once said, "Your readers aren't interested in anything other than information. They usually just read your paper only once, unless they want to repeat some experimental recipe, and then they file it in their minds if it's useful. So scientific papers are a very ephemeral form of literature. People rarely re-read them."[3]

The term "sleeping beauty" has been proposed for papers that are initially not highly cited after publication but are discovered later and become belatedly relevant and cited. Such papers include one by Einstein, which took decades to be noticed (from 1935 until the 1990s), and a paper on adsorption in solutions that was published in 1906 but became noticed in 2002 (the "awakening year").[4]

Generally, scientific papers slot into the knowledge of a field like a brick building up a wall and, as the wall becomes bigger and more solid, less attention is paid to the individual blocks, which once, when the wall was smaller and weaker, stood out more clearly. Francis Crick once suggested that a good test of the historical significance of any work is whether, if it was deleted, it would make any difference to other work.[5]

As an example of the classical march of scientists' findings from vulnerable statement to impregnable fact, few papers or textbooks today feel it necessary to reference the fact that DNA is a double helix to the original report by Watson and Crick in 1953.[6] That paper has nonetheless been cited 7,034 times,[7] and it is useful in exploring the fate of papers to consider the pattern of its citations.

Figure 6.3 shows the number of citations over the years since its publication, both the first two decades or so (main plot) and right to the present day (inset). Figure 6.4 shows the citation map for those papers which cited this paper in the first decade after it was published, with the largest circles representing the most influential (most cited) papers; the multiple pathways by which this discovery percolated rapidly through the literature can in this way be tracked.

To explore further the way in which scientific papers settle into their field after publication, I will discuss two more recent examples, one where a paper was

---

[3] Quoted in Wolpert, Lewis, and Richards, Alison. *A Passion for Science.* (Oxford University Press, 1988), p11.

[4] Cressey, D. "Sleeping beauty" papers slumber for decades. *Nature*, May 26, 2015.

[5] Quoted in Judson, Horace. *The Eighth Day of Creation.* (London: Penguin Books, 1979), p486.

[6] Watson, J. D., and Crick, F. H. C. (1953) Molecular structure of nucleic acids—a structure for deoxyribose nucleic acid. *Nature*, 171, 737–738.

[7] Figure from Web of Science, July 2019.

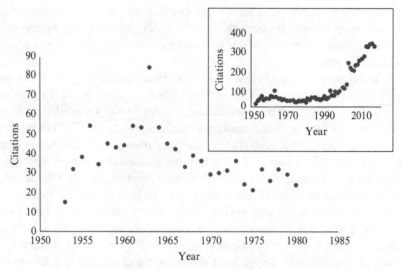

FIG. 6.3 *Citation numbers for the paper by Watson and Crick on the double-helix structure of DNA, from 1954 to 1985 (and over the broader period to the present date, inset). The work generated a lot of initial interest (given the size of the field of molecular biology at the time), and then waned, while interest (measured through citations) increased greatly from the late 1990s onward, due to historical interest in the work.*

FIG. 6.4 *Citation map for the paper by Watson and Crick on the structure of DNA (1962). This map was generated using the VOSviewer software.*

accepted rapidly as being correct and useful, the other where the paper's findings started a firestorm of controversy and argument in the scientific literature. Both were certainly what Kuhn would have described as paradigm shifts, one smashing our long-established belief that carbon came in two forms alone—diamond and graphite—and the other claiming to present evidence that we are not after all alone in the universe, and that our neighbor Mars may have harbored microscopic life, at least, some of which journeyed to Earth millennia ago entrapped in rock.

The former paper was authored by Kroto, Heath, O'Brien, Curl, and Smalley in 1985,[8] and has since accumulated 11,638 citations.[9] The latter had a total of nine authors, will be referred to as McKay et al.,[10] and has been cited 1,129 times since its publication in 1996.[11]

We will consider the paper by Kroto et al. first, starting with a brief description of the work that led to it. For many years, two distinct forms of carbon have been recognized, as any school child could tell you, brilliant priceless diamond and plain old boring graphite. In 1985, however, a small group of scientists discovered a third form, which they christened buckminsterfullerene. While diamond is composed of carbon atoms in three-dimensional lattices, and graphite of carbon atoms in planar flat layers, buckminsterfullerene, in its most common form, is composed of 60 carbon atoms arranged in what resembles very closely a soccer ball. The molecules are roughly spherical, with the atoms being arranged on the surface in alternating hexagon and pentagon shapes, like the patterns formed by the lacings on a soccer ball. While most 20-letter chemical names are unpronounceably constructed from an amalgam of roots of terms relating to chemical structure, designed to baffle rather than illuminate the layman, buckminsterfullerene uniquely was named after a Canadian architect, Richard Buckminster Fuller, who designed dome structures of appearance similar to the structure of the new molecule. Admitting that the name was still a bit of a mouthful, the chemistry community soon came to affectionately call the new molecules "buckyballs."

These highly exotic buckyballs were found in a very mundane place, soot. One of its discoverers, Harry Kroto, was working on carbon chemistry (specifically how chains of carbon atoms could form in space near stars) in Sussex in England, when he became aware of a machine, just constructed at Rice University, in Houston, Texas, in America, that would potentially be useful in his experiments. This machine, a laser supersonic cluster beam apparatus, could vaporize materials through intense heat to produce clusters of atoms or molecules and then analyze these products. The Houston group was led by Richard Smalley; Kroto was

---

[8] Kroto, H. W., Heath, J. R., O'Brien, S. C., Curl, R. F., and Smalley, R. E. (1985) C_{60}: Buckminsterfullerene. *Nature*, 318, 162–163.

[9] Figure from Web of Science, July 2019.

[10] McKay, D. S., Gibson, E. K., Thomas Keprta, K. L., Vali, H., Romanek, C. S., Chillier, X. D. F., Maechling, C. R., and Zare, R. N. (1996) Search for past life on Mars: possible relic biogenic activity in Martian meteorite ALH84001. *Science*, 273, 924–930.

[11] Figure from Web of Science, July 2019.

introduced to a collaborator of Smalley, Bob Curl, while at a conference in Texas,[12] and went with Curl to visit Smalley, where Kroto became excited about the possible applications of the apparatus to his own work. A year later, after some contacts from Kroto, largely through Curl, Smalley invited Kroto to Houston to do a few experiments.

What followed took no more than 10 days, but revolutionized what was known about chemistry and opened up a whole new field of scientific endeavor, with possible applications to everything from materials science to rocket fuel and the fight against AIDS.[13] Kroto designed experiments that two postgraduate students ran on the machine, and in a week they generated reams of data on the type of carbon molecules formed on vaporizing carbon under the intense heat of the laser.

One peak in their data, however, surprised them, and was not easily explained away, corresponding to a compound equal in weight to exactly sixty individual carbon atoms. The team discussed several possible structures for this molecule, none of which appeared to satisfy the unusually specific number of carbon atoms, considering the type of bonds each carbon atom could form with its neighbors. The idea of a spherical structure emerged gradually, until Kroto, an admirer of the architect Buckminster Fuller, suggested a structure with some sort of regular repeating polygon on the surface. Smalley, apparently working with paper, scissors, and sticky tape, managed to make a closed sphere structure by using a combination of 5- and 6-sided rings that had exactly sixty vertices, or edges, where each vertex could correspond to a carbon atom. After some checking, the team agreed that it must be the right structure, their main disbelief being that such an elegant structure was not already known, but it was not. Someone then suggested the name buckminsterfullerene, and thus begat a molecule. The paper describing the discovery was written up feverishly and then, 10 days after arriving, having casually changed the world of carbon chemistry, Kroto flew back to England. In the *Nature* paper that finally announced the discovery, Kroto had first name status, the best seat in the house, while Smalley took last place in the author list, indicating the seniority of his role in the research.

The collaboration allegedly became rather strained after this initial work, over a number of disputes, but the significance of what was achieved in a very short space of time cannot be understated. This is, in my view, a lovely example of how great significant discoveries can just appear. All that is needed is some cosmic conjunction of the right minds with the right ideas being in the right place, at the right time, with the right laser supersonic cluster beam apparatus. The discovery of the buckyball, the significance of which has become more apparent every year since its

---

[12] A lovely illustration of the social and catalytic function of conferences in science, which will be discussed more in chapter 8.

[13] The story of the discovery of buckminsterfullerene is described in detail in:Aldersey-Williams, Hugh. *The Most Beautiful Molecule.* (New York: John Wiley and Sons, Inc., 1995); Taubes, Gary. (1991) The disputed birth of buckyballs. *Science*, 253,1476–1479; and Baggott, Jim. Great balls of carbon. *New Scientist*, July 6, 1991, 34–38.

discovery, was the product of the scientific equivalent of a one-night stand arising from a blind date between people who, briefly connected feverishly by mutual passion for a single fateful experiment, soon cooled to familiar strangers.

What can we learn by studying the citation history for this paper (Fig. 6.5)? For the first several years, citations were relatively slow to accumulate, if very respectable (around 50 per year) for any paper, but around 1990 the paper citation rates increased around 10-fold. This is a very unusual pattern for a paper, and perhaps reflects the sudden emergence of the field of nanoscience (around this time, the prefix *nano-* seemed to suddenly become attached to everything, from medicine to materials). This was a peak that slowly decreased, but citations have increased gradually over the last 20 years or so.

The second example we will consider is the pattern of citations for the paper by McKay et al. on the putative discovery of life in a meteorite, discovered in 1984 in a region of Antarctica called the Allan Hills, that was concluded from initial analysis to be from Mars. The paper was rather controversial on first publication in 1996, claiming as it did that four lines of evidence derived from analysis of meteorite sample ALH84001 indicated that it might have contained alien life. This was a nice illustration of what is called the Sagan Standard (named after the cosmologist Carl Sagan) that "extraordinary claims require extraordinary evidence." The wording used by the authors to justify their radical conclusion is a lovely example of the careful construction of a scientific argument:

None of these observations is in itself conclusive for the existence of past life. Although there are alternative explanations for each of these phenomena

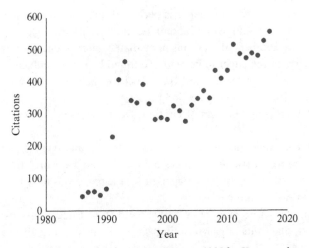

FIG. 6.5 *The pattern of citations for the Nature paper in 1985 by Kroto et al. on the discovery and structure of C60 (buckminsterfullerene). An initial spike of interest was followed by a decrease, but the paper has maintained ongoing (and even increasing) relevance due to its importance in the field of nanotechnology.*

taken individually, when they are considered collectively, particularly in view of their spatial association, we conclude that they are evidence for primitive life on early Mars.

The life proposed to be signaled was microscopic, and indeed one of the initial points of criticism of the work was that the possible cellular structures seen under extremely powerful electron microscopes were too small to be viable forms of life.

On receipt of such a groundbreaking (especially as the ground broken was on a different planet) paper, *Science* sent it to an unusually high number of referees and, on receiving broadly positive reports, published the paper in August 1996.

What followed the publication of this paper was a classical response of a scientific community to a dramatic claim: skepticism and an intense desire to establish how and where exactly the NASA team had made their mistakes. A huge amount of research was undertaken, represented by the huge increase in citations seen in the late 1990s, and most of the published papers presented some form or other of argument as to why the results published by NASA could be explained by other (non-alien-life-based) mechanisms.

Ultimately, it came to be widely believed that the NASA results did not in fact offer conclusive evidence of microscopic life on Mars (and citations started to peter out as researchers moved on to other topics). At the same time, it was widely agreed that the benefits of this surge of interest in life on Mars (from development of new methods, theories, and insights to general increasing awareness of and interest in science broadly) meant that the paper, even if coming ultimately to the wrong conclusions, had a hugely positive impact on its field and beyond.

The citation trend for this paper is perhaps typical of that of many scientific research papers (Fig. 6.6), with different identifiable phases:

Phase 1: low numbers of citations immediately after publication, reflecting the time necessary for the paper to be read and found by others, influence their thinking and direction of research, and then be cited, including the time for a citing paper to work through the publication process;

Phase 2: increasing numbers of citations from papers in the area, likely reaching a peak;

Phase 3: declining numbers of citations, either because (1) the paper is superseded by more recent studies that have built on it, and are more obvious citations than the increasingly old original paper, or (2) the paper was found to be incorrect and not worthy of citation.

The paper by Kroto et al. (Fig. 6.5) showed these three phases (with a longer Phase 1) but, due to its longer-term impact on the field, unusually, never showed a final decline.

Looking at a much older but undoubtedly classic paper, that of Watson and Crick on the structure of DNA (*Nature*, 1954), and focusing only up to 1980, the same three-phase pattern can be seen (Fig. 6.2), while extending the timeline to

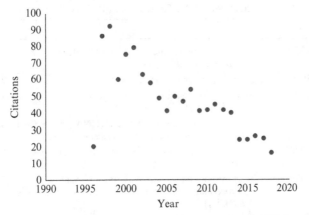

FIG. 6.6 *Evolution of the citation data for the paper by McKay et al. (1996) on possible life on Mars. A rapid surge in interest was followed by a steep decline as the work was first followed up with interest, and then became less relevant as other researchers become increasingly skeptical of the results.*

the current day shows an increase in citations from the 1990s onward, probably driven by ongoing historical interest in the paper and a number of historical anniversaries (such as the 50th anniversary of the discovery, seen as a clear spike in the data).

Megacited Papers and the Dominance of Methodology Papers

The fifteen most cited papers of all time, as listed by an article in *Nature* in 2014,[14] are listed in Table 6.1. One thing is clearly apparent; as discussed earlier, most papers receiving high numbers of citations describe methods that prove useful in the research of others.

Once again, it is also clear that the list is dominated by biology papers, particularly related to protein chemistry and genomic studies, again reflecting the dominance of large-scale biology projects in recent years. The most-cited paper in history is that of Lowry et al. from 1951,[15] which described a rapid and sensitive means of measuring the protein content of solution. The citation trend for this paper, shown in Figure 6.7, demonstrates the same three-phase pattern discussed earlier, but in this case the peak is almost 30 years after the original paper appeared (but probably coinciding with a huge surge in biochemistry research) while the

---

[14] Van Noorden, R., Maher, B., and Nuzzo, R. (2014) The top 100 papers. *Nature*, 514, 550–553.
[15] Lowry, O. H., Rosebrough, N. J., Farr, A. L., and Randall, R. J. (1951) Protein measurement with the Folin Phenol reagent. *Journal of Biological Chemistry*, 193, 265–275.

TABLE 6.1 The Most-Cited Papers of All Time, as of 2019

Rank	Year	Paper	Citations
1	1951	Lowry et al., on the measurement of protein (*Journal of Biological Chemistry*)	305,148
2	1970	Laemmli, on separation of proteins (*Nature*)	213,005
3	1976	Bradford, on measurement of proteins (*Analytical Biochemistry*)	155,530
4	1977	Sanger et al., on the sequencing of DNA (*Proceedings of the National Academy of Sciences of the USA*)	65,335
5	1987	Chomczynski and Sacchi, on the isolation of RNA (*Analytical Biochemistry*)	60,397
6	1979	Towbin et al., on protein analysis and isolation (*Proceedings of the National Academy of Sciences of the USA*)	53,349
7	1988	Lee et al., on an electron density equation (*Physics Review B*)	46,702
8	1993	Becke, on thermochemistry (*Journal of Chemical Physics*)	46,145
9	1957	Folch et al., on isolation of lipids from tissue (*Journal of Biological Chemistry*)	45,131
10	1994	Thompson et al., on genome sequence analysis (*Nucleic Acids Research*)	40,289
11	1958	Kaplan and Meier, on nonparametric estimation from incomplete observations (*Journal of the American Statistics Association*)	38,600
12	1990	Altschul et al., on a genomics analysis tool (*Journal of Molecular Biology*)	38,380
13	2008	Sheldrick, on a short history of SHELX (*Acta Crystallograhia A*)	37,978
14	1997	Altschul et al., on a protein database search tool (*Nucleic Acids Research*)	36,410
15	1962	Murashige and Skoog, on a new medium for tissue cultures (*Physiology Plant*)	36,132

paper continues to be cited right up to the present day, at a rate probably higher than that of the vast majority of more recent papers.

The second most-cited paper is by Laemmli,[16] and describes a method for the separation of proteins on polyacrylamide gels on the basis of their molecular weight, which is a real workhorse technique for protein chemistry laboratories worldwide (in my own work, I have frequently cited both this paper and that of Lowry et al.). The citation trend for this paper (Fig. 6.7) is almost identical to that of Lowry et al., with a peak presumably reflecting peak usage in the field, but an ongoing very respectable rate of citation 50 years after the paper first appeared. Interestingly, the third most-cited paper is also on the measurement of proteins in

---

[16] Laemmli, U. K. (1970) Cleavage of structural proteins during assembly of head of bacteriophage T4. *Nature*, 227, 680.

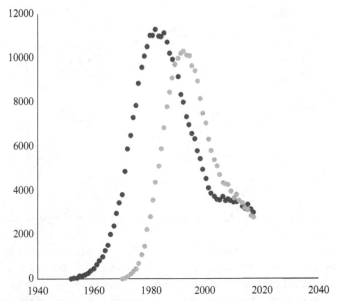

FIG. 6.7 *Citation data for the two most cited papers in history, by Lowry et al. (1951) and Laemmli (1971), both relating to methods for analysis of proteins. The similarity in citation trends is remarkable, as is the fact that both peaked in the 1980s–1990s, despite being published almost 20 years apart, reflecting the huge amount of related research in that period.*

biology (the so-called Bradford method), which has twice as many citations as the fourth most-cited paper, sealing the dominance of protein studies in the scientific literature.

Almost all the remaining top 15 papers seem to involve methodologies, including a paper on mathematical methods (at number 11) and a paper on a method for analyzing crystal structures at number 13, which is, when citations per year are calculated, actually the third most regularly cited paper corrected for age after the first two papers on this list.

From the top of the list of the most-cited papers shown in Table 6.1, it is not immediately obvious that the next most-cited type of paper is a literature review, but this is the case. I have selected one at position 100, with a whopping 12,119 citations, on the topic of the disease atherosclerosis.[17] The citation trend for this paper is shown in Figure 6.8, and a broadly similar "increase-peak-decline" trend is clearly evident, as a paper's usefulness to the field is gradually overcome by the fact that it is missing the most recent developments in the field,

---

[17] Ross, R. (1999) Mechanisms of disease—atherosclerosis—an inflammatory disease. *New England Journal of Medicine*, 340, 115–126.

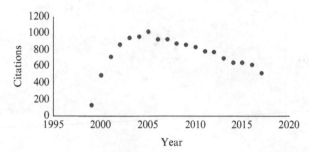

FIG. 6.8 *Citation trends for a highly cited review article, by Ross (1999) on the topic of atherosclerosis.*

and so more recent reviews have probably superseded it as the go-to summary of the state of the field.

This discussion of citation trends actually only considers the least complex form of citation analysis, and far more complex approaches and tools are available for deeper dives into such data. For example, we have just considered first-generation citations, to the initial paper, but it is possible to consider second-generation citations, which are citations to the citing articles themselves. Such analysis gives an indication of the importance and impact of the papers that cited the original paper, but numbers can get rather large rather quickly. For Watson and Crick (1952), for example, the 6,529 first-generation references yield a total of 164,373 secondary citations, indicating that each paper that cited the original *Nature* paper was then cited an average of around 25 times each.

Tracking such secondary citations can add depth to the evaluation of a paper's impact, as papers that are cited by widely cited papers are clearly being recognized by researchers doing impactful work of their own; this is a bit like how Google originally revolutionized Web-searching by use of an algorithm (Pagerank) that ranked search results by the number and quality of links there were to the sites found.

Broader citation analysis has even proven useful in identifying so-called citation cartels, where researchers collude to improve their research metrics in a "you cite me and I'll cite you" arrangement, where researchers cite each other disproportionately more than other studies in the same field that in theory could be cited instead.

Finally, returning to the case of the discovery of gravitational waves described in chapter 4 gives us a final example, appropriately, of the ripples even a recent paper (two years old) can cause in a scientific field when it represents such a critical development, in that case the confirmation of an almost century-old prediction. Despite the short time elapsed since publication, citation maps can be drawn (Fig. 6.9), which show the number of papers citing and tracing networks and relationships in the physics community (and note this is only for the first 500 papers published that cited the discovery).

FIG. 6.9 *Citation map, generated using VOSviewer software, for the paper by Abbott et al. (2016) on the discovery of gravitational waves.*

The Ongoing Evaluation of Papers: Postpublication Peer Review

There are now a number of websites dedicated to the hosting and discussion or analysis of papers following publication. Perhaps the best known of these is PubPeer, which labels itself an online Journal Club,[18] where researchers can critique, comment on, or criticize published papers, with authors being notified when comments about their paper are posted. This gives papers an ongoing forum as a seed for discussion, interaction, and stimulation of developments in the field, and has also led to a number of cases where papers that attracted a lot of negative or suspicious discussion eventually were corrected or even retracted. This is a form of crowdsourced real-time application of the principle that one of the greatest strengths of science is that it is ultimately self-correcting, and that errors (or worse) eventually get discovered and fixed one way or another, without needing to wait for later follow-up papers to point out flaws or irreproducibility of published results.

---

[18] A journal club traditionally is a research group practice where members of a team find and discuss interesting papers from relevant literature.

ResearchGate is described as a social networking site for researchers, whereby they can share their papers (or data, patents, or other research-related documents) with peers, engage in discussion around their work, and perhaps find collaborators. The site has been somewhat controversial for hosting papers that are in infringement of copyright, and it is stressed that only copyright-free prepublication authors' final draft manuscripts can be uploaded. I have a ResearchGate account myself, which I rarely proactively manage or use, but which still seems to generate a lot of email traffic in terms of requests for articles, notifications of uploads, and notes of congratulations on reaching some milestone or other (such as reaching a particular number of reads). On checking the site, a huge number of my publications seem to be listed, and I apparently have a ResearchGate Score higher than that of 97.8% of ResearchGate members, which at least sounds impressive!

In this chapter, we have seen how papers, in practical terms, create impact and add to the body of knowledge in their field, by being taken up and used (and then cited) by others. Each published paper is a nucleus, which sometimes leads to nothing (most researchers will have some papers they would have predicted to be well cited, but to which their peers were apparently indifferent) but sometimes becomes the seed of a blossoming network of influence, impact, and citation, spreading rapidly across the globe as a messenger of information, and hopefully ending up yielding the best desired outcome for that particular piece of work.

Of course, when information can today circulate the globe so rapidly and be rapidly taken up by others, and influence their actions, a key assumption is that the information presented is true, and is a fair basis for others to trust and work on. Is this not always the case?

{ 7 }

Ethics and Integrity in Scientific Communication

The *Oxford English Dictionary* (OED) defines ethics as "Moral principles that govern a person's behaviour or the conducting of an activity . . . the moral correctness of specified conduct . . . the branch of knowledge that deals with moral principles." Ethics relates to what is morally considered right and wrong.

For many years, the "right" way to undertake scientific research and publication seemed pretty obvious; people would publish only genuine results in an honest manner, and act professionally toward their colleagues, peers, and discipline. Scientists were assumed to be honest and noble (well, there had to be some explanation for suffering the long hours, poor pay, and frequent disappointments), interested solely in the truth. Artifice and dishonesty seemed as distant from the ethos of scientific writing as one could imagine, more associated with criminals or perhaps politicians.

One Would Have Imagined Wrong, Unfortunately.

I am sure there is some apt biblical analogy for the fall from grace of science since the mid-1980s, as the acronym FFP (fabrication, falsification, and plagiarism) increasingly entered the scientific and broader lexicon, and we arrived at a place where scientists are made aware from the start of their careers of the ethical requirements and integrity considerations associated with their chosen career.

Over the last 30 years or so, there have been several very high-profile cases of clear misconduct in research, from almost every field of science, and these have attracted negative publicity from the media, funding agencies, and governments, and rightly so. Estimates today of the prevalence of misconduct vary, as we will see, depending on the definition of misconduct applied, but inevitably the high-profile cases throw disproportionately long shadows.

Dispassionate observers may conclude that it is an easy defense for scientists to claim that misconduct and fraud are rare in science, given the evidence, but it is surely fair to give the benefit of the doubt to the majority of scientists, who abhor the behavior of those who besmirch their profession. The key today is recognized as ingraining in young scientists the importance of honesty and the consequences of dishonesty, and then implementing systems that, among other goals, try and determine the veracity of results submitted for publication without rendering the process of publication unworkable.

Today, the broader concept of research integrity has come to encompass what is regarded as acceptable behavior for researchers, across all aspects of their activities, and research ethics is sometimes defined as a subset of research integrity. Integrity can be regarded as relating to professional standards and codes of conduct, while ethics relates to moral considerations associated with specific types of research.

Some areas of research come with detailed ethical codes and are tightly regulated due to the potential for morally questionable actions or consequences. This might include research on human subjects, where the risks and benefits of the research, and personal implications for the well-being of participants, need to be explicitly articulated to potential participants, who only then choose of their own free will to give their consent to be involved. Researchers planning such studies will have to get advance permission to conduct their studies from institutional ethics boards, outlining their plans and the manner in which safety, dignity, and privacy will be protected, and when later submitting their research for publication will be required to include proof that such approval was correctly granted. Such safeguards will also be in place for research involving animals, to ensure that every precaution is taken to avoid unnecessary pain or risk, or social science studies where confidential information is obtained from subjects through surveys or interviews.

In my own work, ethical approval is not needed to obtain milk from cows or farms but, when working with human milk, it most certainly is, and, if a change to a study is proposed midway, even to include a type of analysis not mentioned in the original application for ethical approval, this also must be approved by the ethics board.

So, some types of research are carefully regulated to ensure that no morally unfit practices (such as abusing patients or infringing their dignity, using their personal details in a manner to which they have not agreed, or exposing more animals to a test treatment in a veterinary trial than are necessary) can occur. On the other hand, many other types of research, for example in chemistry, physics, or indeed food science, do not involve such considerations, due to their lack of such directly moral considerations. Some researchers will thus be working in more ethically complex areas than others.

However, every single researcher is subject to expectations around research integrity, requiring that they conduct themselves and their work according to the highest professional standards, including honesty and responsibility.

Why Would Scientists Be Dishonest?

Taking as a starting point for discussion the key tenet of integrity that states the need for researchers to be honest, two questions that follow are how and why researchers could be dishonest.

Going back to the OED, the word "honest" is defined as "free of deceit: truthful and sincere."

A researcher could be dishonest at almost every stage of the research process, from proposing their research ("We are able [when actually we are not] to conduct this research due to our [nonexistent] active research program in . . . ") and undertaking it ("This won't hurt" or "that result looks odd so let's discard it") to reporting it, which will be the focus of this chapter, in keeping with the themes of this book.

If we go back to key points discussed previously, one of the most important things for a researcher is their publication record, which profoundly influences their career options, and hence life direction. The better that record, the more personal benefit they will get and, as in every way of life, this creates temptations for individuals to take steps which will increase that benefit, perhaps because their paper gets into a better journal, or they get more papers, or they gain numerous other forms of advantage.

The bank robber Willie Sutton, when asked why he kept robbing banks, is alleged to have said, "because that's where the money is." In research, a key place where reward is found is through publication, which is why honesty in publication is a key focus of all modern frameworks for research integrity. There can be no "fake news" in scientific publishing.

The American psychiatrist Donald Kornfeld, who has written extensively about research misconduct, categorized several drivers of individual misconduct, including the following:[1]

- Desperation, where fear of failure overrides other codes of conduct;
- Ego/grandiosity, due to someone's belief in their own intelligence or superiority;
- Ethical challenges, where someone's personal moral framework is not adequately robust to lead them not into temptation toward inappropriate conduct;
- Sociopathy, where a person's lack of a conscience is manifest in research misconduct, rather than more traditional forms of crime or violence;
- Lack of proper initiation, where a young researcher is not properly mentored or trained in appropriate scientific behavior;
- Perfectionism, where an individual regards any form of failure as unacceptable.

These are all human factors that could just as easily apply to any other form of misbehavior in which people engage outside the realm of science, which reminds us that human elements and behavior cannot be discounted when considering, and coming up with systems to protect, the edifice of scientific research. In addition,

---

[1] Kornfeld, D. S. (2012) Perspective: research misconduct: the search for a remedy. *Academic Medicine*, 87(7), 877–882.

institutional systems can indirectly incentivize bad behavior, through pressures
applied to get promotion, advancement, or credit.

Defining the Main Issues

Any international code of practice around research integrity typically refers to
FFP. Fabrication and falsification differ in that the latter refers to the inappropriate manipulation of data (such as omission of inconvenient data points), so that
a figure or table that appears in a paper is not an open and honest depiction of
exactly what the results were, whereas fabrication refers to the complete invention
of data, such that a statement, figure, or table is wholly or partly based on data that
have no basis whatsoever in reality. Plagiarism then refers to the use of the words
or ideas of another without allocation of due credit.

In terms of a broader systematic classification of types of misconduct, the
OECD (Organisation for Economic Co-Operation and Development) classified
the following in a 2008 report:[2]

1. Core research misconduct (FFP);
2. Research practice misconduct (inappropriate methodology, poor research
   design);
3. Data-related misconduct (bad data management, storage, withholding data
   from the community);
4. Publication-related misconduct (inappropriate authorship, "salami-slicing"
   [artificially proliferating publications], failure to correct the publication
   record);
5. Personal misconduct in the research setting (inappropriate personal
   behavior, harassment, poor mentoring, social or cultural insensitivity);
6. Financial and other misconduct (peer-review abuse, lying on a CV, misuse
   of research funds for personal gain, making unsubstantiated or malicious
   misconduct allegations).

How Big Is the Problem?

The answer to this question can frequently depend on the person being asked.
Researchers will instinctively assert that the vast majority of scientists are of
course honest. Honesty in scientific writing is assumed, and fraud is barely con-
templated. Carl Djerassi, a chemist and occasional novelist, said, "Fiction is
almost the antithesis of science. Scientists never have the luxury of being able to

---

[2] Organisation for Economic Co-Operation and Development Global Science Forum. Best
Practices for Ensuring Scientific Integrity and Preventing Misconduct (https://www.oecd.org/sti/sci-
tech/40188303.pdf).

say 'ah, but I made it up'. That's *verboten*. So fiction is a wonderful luxury. You can even brag about it. When you say 'I made it up' people congratulate you! Scientists would kick you out; they would say you are finished."[3]

On the other hand, the media, politicians, regulators, and those with a personal interest in the corruptness of science might give very different perspectives. For these reasons, the most reliable sources of information on frequency of scientific misconduct are likely to be confidential surveys, where researchers are asked about their own conduct, or their awareness of the conduct of colleagues.

In an early systematic study on this, a report in *Nature* in 2005 on a survey of 3,247 biological scientists[4] included the following headline findings:

1. Falsifying research data—0.3%;
2. Using another's ideas without obtaining due permission or giving due credit (plagiarism)—1.4%;
3. Publishing the same data in two or more publications—4.7%;
4. Failing to present data that contradict one's own previous research—6.0%;
5. Inappropriately assigning authorship credit—10.0%;
6. Withholding details of methodology or results in papers or proposals—10.8%;
7. Overlooking others' use of flawed data or questionable interpretation of data—12.5%;
8. Dropping observations or data points based on a gut feeling they were inadequate—15.3%;
9. Changing the design, methodology, or results of a study in response to pressures from a funding source—15.5%.

While the incidence of the most clearly inappropriate behaviors was low here (although a large sample size means that 10 respondents admitted to falsification, for example), there is a rather high level of reporting of what has come to be termed "questionable research practice" (QRP). Interestingly, respondents were subdivided into early- and midcareer researchers and, of the list above, issues 3, 5, and 6 were reported significantly more widely by the latter than the former. Overall, one-third of respondents admitted to participating in at least one of the forms of behavior listed in the survey (not all of which are shown here) in the previous three years.

A 2008 study by the US Office of Research Integrity, which oversees misconduct reporting and cases in health-related research, reported that around 8.7% of researchers (out of a sample size of 2,212) had observed instances of likely misconduct over the previous three years.[5] When the responses were analyzed to

---

[3] In an interview in Wolpert, Lewis, and Richards, Alison. *Passionate Minds: The Inner World of Scientists.* (Oxford University Press, 1997), p11.

[4] Martinson, B. C., Andersen, M. S., and de Vries, R. (2005) Scientists behaving badly. *Nature*, 435, 737–738.

[5] Titus, S. L., Wells, J. A., and Rhoades, L. J. (2008) Repairing research integrity. *Nature*, 453, 980–981.

see what fit a definition of FFP in proposing, performing, or reviewing research or in reporting research results, 7.4% of reports fit this criterion, splitting two-thirds for fabrication and falsification and one-third for plagiarism. The authors made several recommendations as to how to improve a culture of integrity, including adopting a zero-tolerance policy toward misconduct, protecting whistle-blowers, and improving training of mentors of young researchers.

There have been several other studies, some in other fields, from economics to psychology, that have reported broadly similar trends and incidence levels. A meta-analysis from 2009 of 18 surveys, data from which were crunched together and blended to extract a purified essence of common themes and outcomes,[6] concluded that just under 2% of scientists across a wide range of fields had fabricated, falsified, or modified data or results at least once, while up to 34% had admitted other forms of QRP. When asked about the behavior of colleagues, however, the results for fabrication and falsification increased to 14% and that for QRP to 72%.[7] The analysis also suggested higher levels of misconduct in medical research than in other fields, but it was noted that this may reflect higher awareness of standards and likelihood of reporting in that field.

A later survey in the UK of 187 researchers in the biological sciences[8] found that inappropriate allocation of authorship was the most prevalent inappropriate behavior, while fabrication and plagiarism were not determined to be prevalent, although another meta-analysis published in 2015[9] concluded that 1.7% and 30% of scientists admitted to committing or witnessing plagiarism, respectively, and that these rates were higher than, but correlated with, rates of fabrication and falsification. On a positive note, the latter analysis concluded that the incidence of FFP reported in surveys has declined over time.

Nonetheless, a rather unsettling observation from Richard Horton, retired editor-in-chief of the medical journal *The Lancet*, in a commentary in that journal in 2015, was that "the case against science is straightforward; much of the scientific (biomedical) literature, maybe half, may simply be untrue," based on his seeing many studies with small sample sizes, tiny effects, flagrant conflicts of interest, and a tendency to follow "fashionable trends of dubious importance."

---

[6] Fanelli, D. (2009) How many scientists fabricate and falsify research? A systematic review and meta-analysis of survey data. *PLOS-One*, 4(5), e5738.

[7] This may reflect a similar phenomenon to that which explains the fact that the poll that most accurately predicted the outcome of the 2016 US election did not ask those polled who they would vote for, but rather who they thought their neighbor would vote for. People might be drawn to projecting their own wishes or actions, of which they are ashamed, onto others as a way of getting them off their chests without admitting an unpalatable truth to themselves. Something called the "Mohammed Ali effect," whereby people tend to perceive themselves as more honest (but perhaps not more intelligent) than their peers, may also come into play here.

[8] Roberts, D. L., and St John, Freya, A. V. (2014) Estimating the prevalence of researcher misconduct: a study of UK academics within biological sciences. *Peer J*, 10.7717/peerj.562.

[9] Pupovac, V., and Fanelli, D. (2015) Scientists admitting to plagiarism: a meta-analysis of surveys. *Science and Engineering Ethics*, 21, 1331–1352.

Publication and Retraction

Any scientific publication is a permanent entry in the record of human knowledge, and will be assumed to be a legitimate and honest report of work done and outcomes recorded. It has to be this way, as for researchers to approach every paper they read with a skeptical eye and suspicion would make the transmission of knowledge and the progressive developments of fields far more difficult, as well as ultimately eroding trust and morale in the enterprise of research.

However, there then needs to be a process by which that which is found to not meet the standards expected is labeled accordingly and warnings erected so that readers either ignore that paper or at the very least approach it well warned as to issues that might impinge on their interpretation of what they read.

This process is called retraction, meaning the withdrawal of a statement or assertion. When a paper is retracted, the journal has, based on evidence that the editors and publishers believe to be valid and unarguable, labeled the paper in whole or in part with an actual water-marked digital stamp stating "retracted" (the equivalent to "warning: approach with caution"), accompanied by a note indicating the reason for the retraction. In the pre-Internet dark ages (i.e., pre-1990s) such a process would be rather difficult, unless someone went around every library and subscriber's office with an ink stamp or scissors, and linking later published notices of caution of retraction to previously published papers was difficult. Today, with online publication, this has become technically far easier and, once a published paper is retracted, the ripples of evidence of its existence (on databases, etc.) will be similarly corrected.

The first reason why the retraction may have occurred is the most benign and favorable, which is where authors have discovered, after publication, a problem or mistake in their paper, which makes it incorrect or unreliable, and they need to correct the scientific record accordingly. They have contacted the journal to request the retraction, in line with the expectations of researchers acting with the appropriate level of scientific integrity, and they will suffer no indignity or disrespect as a result. While doing such is undoubtedly embarrassing, the responsibility lies with the authors to take this step, and they will be lauded by their peers for doing so and, it has been reported, not suffer any impact on their own citations (compared to those for whom retraction is involuntary).[10]

Indeed, I am sure there are cases of later problems that manifested as serious issues of misconduct that started out with someone discovering a mistake they had made and deciding not to retract the paper, perhaps hoping that no one would notice, but this became the start of a slippery slope.

The more problematic retractions, then, are those which are not voluntarily initiated by the authors of a paper but are the decisions of editors and publishers

---

[10] Lu, S. F., Jin, G. Z.,Uzzi, B., and Jones, B. (2013) The retraction penalty: evidence from the web of science. *Scientific Reports*, 3, article number 3146.

based on information they have received from other parties, who may be readers, other researchers, or perhaps whistle-blowers with direct knowledge of the authors' practices or issues with the paper.

I have said repeatedly that one of the most precious things to any researcher is their reputation. Retraction (other than the voluntary self-driven kind), where one's online evidence of success through publication is visibly and publicly blemished, is one of the most damaging (depending on circumstances and severity) blows to reputation. Of course, a key question, then, concerns other repercussions of the misconduct that has led to the retraction, which we will discuss later. For now, we can note that retraction is the scientific equivalent of "naming and shaming," particularly when a researcher's name appears on the rogue's gallery of shameful incidents reported on the excellent website *Retraction Watch*.[11]

We will next consider briefly some case studies of extreme cases of misconduct leading to retraction.

Fabrication and Falsification

The history of science is sadly replete with instructive examples and teachable moments where researchers have behaved in a manner that is clearly in breach of the principles of good conduct outlined previously. Indeed, one of the first real articulations of the problems of fraud in science was by Charles Babbage (1791–1871), the British mathematician who is regarded as one of the fathers of computer science, in a book called *Reflections on the Decline of Science in England, and Some of Its Causes*, published in 1830.

In that book, Babbage spoke of fraudsters who allow prejudices and biases to influence their observations and conclusions. He also defined several types of scientific fraud, including hoaxing (deliberate planting of false evidence to test credulity), trimming (eliminating some data points to make results look more accurate), cooking (selecting data in a less than transparent manner), and forging (what is today called fabrication, which is designed not to be detected, unlike hoaxing, which generally is).

In this section, I will discuss a small number of high-profile cases, which are chosen because of their illustration of key concerns or factors, as well as the response of the scientific community.[12]

---

[11] Retraction Watch (retractionwatch.com) was launched in 2010 with the mission of tracking retractions as a window into the scientific process, and improving the transparency of the retraction process. With its pithy summaries of retractions and much broader range of article on scientific conduct and misconduct, I couldn't recommend a better resource for anyone looking to learn more about science and scientific standards than to become a regular reader or subscriber to their alerts to new postings.

[12] A very good, if rather pessimistic, account of several of these cases and many more can be found in a very thought-provoking book as follows: Judson, Horace Freeland. *The Great Betrayal: Fraud in Science*. (Harcourt Inc., 2004).

As a starter, in the 1980s a researcher called Elias Alsabti (from either Iraq or Jordan, accounts differ) was found to have falsified his entire curriculum vitae and embarked on a process of taking papers from obscure journals and resubmitting them in his own name to different journals; he even actually stole papers submitted for review for other journals, which he found in the mailboxes of colleagues in his university. When the net started to close in on him, he apparently disappeared, in a twist that reminds me of the character Keyser Soze in the film *The Usual Suspects.*[13]

An infamous case that perhaps illustrates clearly some of the pressures and factors that lead to conduct that is simply unacceptable in research is that of Jan Hendrick Schön, a German researcher at the renowned physics research center at Bell Laboratories, who was studying means of making transistors, which are key components of computer chips, in the late 1990s and early 2000s.

Schön published a huge number of papers in the late 1990s and early 2000s (at one stage, one paper appeared under his name every eight days) in the top journals in the world, such as *Nature* and *Science*, and won several awards for his work. However, his Web of Science records today show 42 out of 98 listed papers as being retracted, a very dramatic fall from grace, and a stark example of expunging a significant body of invalid work from the literature.[14]

The initial concerns about his work were raised by readers of his papers, when no one was able to reproduce his findings of spectacular performance for a very new type of transistor based on organic compounds. This is a lovely example of the self-policing nature of science, where apparently significant research outcomes will be carefully scrutinized by the relevant research community, and failures of reproducibility or other perceived problems (including anomalies in his data, such as curves appearing far too similar, even down to what should be random noise, and the same curves appearing in multiple papers claiming in each case to be different things) will rapidly draw negative attention to a paper (or, in this case, a whole lot of papers). Schön was unable to refute the allegations because he said that many of his claimed devices had been damaged or destroyed, and much of his data was not recoverable as he had to delete it from a computer that apparently had insufficient memory.

A further controversial aspect of the Schön case was that he had several coauthors across the retracted papers, who were all exonerated of scientific misconduct, which raised the issue of culpability of coauthors of fraudulent papers, and the extent to which one bears responsibility for any paper to which your name is attached (in other words, just as researchers reap benefits from successful and highly recognized papers, should they not share blame when the opposite happens?). Another example of how these high-profile cases influenced future developments of codes of good practice for research is that they influenced stringent

---

[13] He did resurface and later tried to practice as a doctor, but had his medical licence in the United States suspended a number of times. He died in a car accident in South Africa in 1990.

[14] Schön's PhD was also later revoked on the basis of "dishonorable conduct."

expectations today as to how researchers should maintain and archive research results for future inspection if required.

As with any crime or other offense, an obvious question afterward is why the perpetrator did it in the first place. In the case of Schön, the need to continue to perform and live up the expectations generated by his first (genuine) papers appears to have been a significant factor, showing the importance of ego and the dangers of a need of researchers to live up to a reputation and impress others, perhaps due to their own need for personal affirmation. He also seemed to determine from talking to colleagues what results were hoped for, and then worked backward, as it were, to deliver these. As discussed throughout this book, the practice of publication is intrinsically linked to a researcher's own career development and future opportunities, and this could be a clear example of how such pressures drive some individuals to behavior that is clearly in breach of scientific integrity.[15]

The Schön case also illustrates how factors that might be imagined to reflect positively on researchers, such as publishing a lot of papers in high-impact journals, with such papers containing very nice clear results and plots, can actually perversely raise red flags regarding breaches of research integrity and inappropriate conduct. There is simply a limit to how many papers a researcher can realistically produce in a given time period, based, among other factors, on the time required to generate data and the number of hours in a day. Excessive levels of publication might in this light be suggestive either of honorary authorship (where someone's name appears on papers to which their actual contribution of effort and time was none to minimal) or outright fabrication of data generated without any labor and time being expended.

In terms of examples of researchers breaking the "rules" of research, I sometimes think that if I sat down and tried to invent a case study that showed every single thing that scientists should not do, I couldn't do better than the Korean stem cell scandal, which focused on the claim in 2004 of the team of Woo-Suk Hwang from Seoul National University in South Korea to have cloned embryonic stem cells. The offenses perpetrated covered the spectrum from bioethics law violations (coercing lab workers to donate eggs without proper informed consent, theft of cell lines from hospitals) to financial fraud, political interference (as the Korean government tried to close down investigations into what was at the time the greatest scientific success story to come from that country), and publication misconduct, which is the only aspect I will focus on here (as the whole story could fill a book by itself[16]).

In the mid-2000s Hwang and his team (including one American coauthor) published in *Science* several papers on their discovery, which have now been retracted

---

[15] The Schön case is detailed in a highly readable manner in the book: Reich, Eugenie. *Plastic Fantastic: How the Biggest Fraud in Physics Shook the Scientific World*. (New York: Palgrave Macmillan, 2009).

[16] One book to explore the topic is: Harold, Eve. *Stem Cell Wars: Inside Stories from the Frontlines*. (Palgrave Macmillan, London, 2006).

for significant problems, including falsification and duplication of images of a small number of cells to make then appear to show things they did not. In addition, the work had not been conducted as claimed, and key results had been faked to show the desired results. The American coauthor also admitted to having a minimal contribution to the work (certainly less than would warrant authorship), and it was suggested that his main function was to give the work a higher chance of being published in Western scientific journals; he was ultimately deemed to have committed "research misbehavior" by not fulfilling his responsibilities as senior author to verify the authenticity of data.

Recommended changes in journal practice that resulted from this scandal included greater scrutiny for high-visibility papers, requirements for statements of contributions of individual coauthors, greater availability of raw data on which papers are based, and agreement of consistent reviewing practice between top journals, so that none could be perceived as being a somehow easier target for authors seeking to publish fraudulent papers.

Hwang claimed he was blinded by work and his drive for "achievement," and sought to blame other members of his own team for deceiving him, but his work is now completely discredited and his reputation (and that of all others involved in his work) is in tatters.[17] Penalties and charges in this case, however, went far beyond reputational damage, with accusations against multiple members of his team including fraud in procuring government grants, bioethics law violations, destroying evidence, and obstructing research work (six members of his team were suspended from the university in 2006, and Hwang was sentenced in 2009 to a two-year suspended prison sentence, which was later reduced on appeal).

In another extraordinary case, a paper published in *The Lancet* in 2005 focused on the relationship between nonsteroidal anti-inflammatory drugs (such as household staples like aspirin and acetaminophen) and their potential for reducing the risk of development of oral cancer. After publication, it was discovered that none of the alleged 908 participants (equal numbers of subjects and matched controls) actually existed (250 of them apparently had the same birthday), and the authors (which included, again bizarrely, the first author's wife and his identical twin—leading possibly to a defense of "my evil twin made me do it") had fabricated the entire study. This case also raises, to me, the question of whether the field of a paper, and its impact on peoples' lives, has some bearing on the significance of fraud, and potentially the consequences of such behavior.

Another key question arising from such cases of outright fraud is the ability of peer review to detect deliberate attempts to concoct data, which is an ongoing matter of discussion. It has been reported that it is very difficult to make faked data look "natural," as humans find it difficult if not impossible to make the digit

---

[17] The case remains so toxic that there were huge protests in South Korea in 2017 when a coauthor on one of the papers, who was deemed to have been an influential supporter and enabler of his work, was appointed head of a new science center in the country.

distributions in sets of data resemble those that nature will create, which have certain inherent characteristics of randomness and uniformity. Apparently, we all have subconscious biases toward certain digits, and there are principles that describe the digit distributions in "natural data" that are hard to mimic.[18]

There have been a number of attempts to develop software systems for the detection of scientific fraud, and studies using visual detection to determine the prevalence of image duplication, repositioning, alteration, cutting, or "beautification." In a 2016 study,[19] over 20,000 papers were visually screened for evidence of such treatments, and 3.8% of these papers were concluded to show such manipulation, which appeared to be deliberate in around half of these cases (and showed evidence of repeated bad behavior by a number of authors).

A number of individuals have become quite specialized in the "policing" of the scientific literature, and these are sometimes referred to (in some case, by themselves) as "data thugs." A British anesthetist, John Carlisle, played a key role in pointing out significant problems with data published by a Japanese anesthesiologist named Yoshitaka Fuiji, when he was challenged by a journal editor to back up concerns he expressed in a letter to that journal. Carlisle's approach (which has been subsequently used in other such investigations) involved comparing some of Fuiji's data to that reported in other studies and to that which would be expected to be found by random chance, leading to the conclusion of there being a vanishingly small likelihood of the data being genuine; this was part of the case that ultimately led to 171 papers (!) being retracted on the basis of data fabrication.

While outright fabrication of data can be defined and (hopefully) detected, the determination and definition of falsification, where real data are manipulated in a manner that is less than honest or acceptable, can be more tricky. There have been a rather surprising number of cases where the work of key figures of the history of science has been retrospectively analyzed using modern standards and perspectives and occasionally found to be less than impressive.

For example, Louis Pasteur has been (controversially, of course) accused of being less than transparent about some of his work on rabies vaccination; this includes his apparently claiming to have undertaken more experiments than he had and reversing the order of work in his laboratory notebooks when he came to publish it, as well as acting unethically in proceeding too rapidly from studies on a small number of animals to testing his vaccine on an infected child. Similar questions have been raised about the classical work on genetics of pea plants of Gregor Mendel, as it has been claimed that statistical analysis showed atypical distributions within his data.

One historical example is worth considering in a little more detail, as it very clearly illustrates the complexity in defining falsification, and the very fuzzy border

---

[18] Including one called Benford's Law, which basically states that first digits in real numbers tend to be small, and the likelihood of a digit appearing decreases from 1 to 9.

[19] Bik, E. M., Casadevall, A., and Fang, F. C. (2016) The prevalence of inappropriate image duplication in biomedical research publications. Available currently as a preprint on bioArxiv.

between, on the one hand, good scientific practice and the use by researchers of intuition to know how to trust and understand their methods and equipment, and, on the other, bias and inappropriate manipulation of data to yield a desired result. This relates to the Nobel Prize–winning and painstaking work of the physicist Robert Millikan to measure the charge on the electron, which he performed by measuring the rate at which oil droplets, bearing different levels of electrical charge, fell through a chamber. The rate of fall was proportional to the charge and, by collecting a set of values for different droplets, he was able to calculate the basic unit of charge. In his classic paper published in *Physical Review* in 1913, he reported that he had studied 58 different oil drops, and used the much analyzed wording "this is not a selected group of drops but represents all of the drops experimented on during 60 consecutive days," leading to measurements on 58 drops "with no single drop being omitted." However, later analysis of Millikan's notebooks (particularly by a historian called Gerard Holton) showed data for around 140 drops, and beside many of his measurements he had written notes like "almost exactly right and the best one I ever had" and "perfect publish," but also "very low, won't work out" and "error high, will not use."

While these notes beautifully capture the real-time excitement of his work and discovery, in a manner too frequently expunged from final published papers, the key issue is whether his wording in the paper, so precise in its statement that all drops measured were used, was disingenuous in terms of not representing honestly what he did. When the charge was calculated using the full data set, the same value was found, albeit with a larger error, but it is clear that Millikan disregarded values obtained in early tests before he believed his apparatus was working optimally, and later ones when it was clear something was not working right. It is clear that such measures are part of normal science and indeed good practice by an experienced experimentalist, but the issue remains that the wording in the paper could be interpreted as not reflecting his true process of data collection, which is why this remains a widely used cautionary tale in science.

The Plague of Plagiarism

Plagiarism can be defined as someone representing the work of someone else as their own, where the work could be not only text (the typical understanding, where someone cuts and pastes text to claim it as their own) but also ideas. Today, in schools and universities around the world, submitted work is routinely electronically screened for textual overlap with previously published work, and commission of plagiarism is regarded as a serious academic offense.

It is not just students who are capable of such deeds though, and plagiarism is a sufficiently well recognized problem in the world of research for it to represent one of the three legs of the FFP triad. Papers submitted to journals sometimes include whole chunks of other work being reused without credit or attribution, a problem that might be encountered proportionately most frequently in review articles, which might be stitched together from sections of text lifted from other papers. When the journal of which I am an editor first instigated random checks of submitted papers through plagiarism-detection software, we were shocked by the level detected, with some papers having up to 70% of their text being determined by text-comparison software as nonoriginal.[20]

We were also surprised and disappointed by the fact that so much of what was detected showed overlap not with the work of other authors but rather with previous papers by the same group. Authors were reusing parts of their own Introduction, Discussion, and (perhaps most inevitably and, in my view, least obviously problematic if not done to excess) Materials and Methods sections. Sometimes this is referred to as "text recycling," which makes it seem as if it is a practice done for environmental reasons, but we, and I suspect many journal editors, take the position that all text must be original when submitted as part of a new research paper.

I guess I am fundamentally an optimist, in which spirit I believe plagiarism is a problem that will diminish with time, perhaps more so than other kinds of research misconduct. The reason for this assertion is that, although developments in computer usage in research made plagiarism much easier than ever before (cut-and-paste rather than retyping), the evolution of computer software to match text, backed up by ever more powerful search engines and databases, quickly made detecting such offenses a trivial matter. Then, when individuals are caught (either before or, worse, after publication) the embarrassment of the admonition should, you might hope, encourage them to avoid taking such risks again.

As an example of the embarrassment that can ensue, a paper was published in a journal called *Comprehensive Reviews in Food Science and Food Safety* a couple of years ago, which is now accompanied, like a little dark cloud, by a published Erratum pointing out that it had been noted after publication that the review contained significant portions of text that had appeared previously in 12 other articles and that the authors apologized to the authors of those papers for this "error." The Erratum also takes the opportunity for the publishers to point out authors' responsibilities in the area of originality of text and plagiarism. This is undoubtedly embarrassing for the authors of the review thus branded, as any search of a database for their work will also turn up the Erratum. With increasingly universal use of text-comparison software for submitted papers, it is unlikely that such cases of postpublication detection will be common in the future, and authors' blushes (and worse) can be handled out of view of their peers.

[20] In a very different field, it was reported in 2016 that 17% of 400 submissions to a major medical journal had unacceptable levels of publication. Higgins, J. R., Lin, F.-C., and Evans, J. P. (2016) Plagiarism in submitted manuscripts: incidence, characteristics and optimisation of screening—case study in a major specialty medical journal. *Research Integrity and Peer Review.* 10.1186/s41073-016-0021-8.

I once found myself the victim of an attempted plagiarism, as it were, when I was asked to review a paper (submitted to a journal we will call A) on a particular preservation technology on oysters, a topic closely related to work one of my former PhD students had completed relatively recently. The student, now graduated, was still working in my university and so (after notifying the editor) we agreed to review the paper collaboratively and arranged to meet on a certain day to agree our report. I vividly remember the day before our meeting sitting at home to read the paper and getting the strangest chill when I read the words "the oysters were scrubbed under fresh running water to remove mud and other adhering debris"; this déjà vu was because I had suggested those words myself for one of my student's papers and now they were appearing on a paper from a group most of the way around the planet.

On reading closely, it turned out that the authors had clearly read and liked our paper, and then repeated the study, with some tweaks and new treatments and analyses. The Introduction to the paper, however, was more or less ours, as was most of the Materials and Methods and much of the Discussion. The figures and tables were even styled after ours, albeit with new data!

So I rejected the paper, explaining why (while trying to make it clear I was not one of the authors of the original paper) and telling the authors that the paper would have to be completely rewritten to make it original, and I separately made the problems clear to the editor as well. They rejected the paper.

Then, a few weeks later, I got a request from journal B to review a similarly titled paper, which I accepted to discover it was the same rejected paper resubmitted without taking my advice on board. So, I rejected it again, whereupon the editor of journal B told me the authors had actually submitted two papers, and when they showed me their second paper it was closely lifted from one of our other papers, so I got to reject that one too.

This was around 8 years ago and, a couple of years ago, after mentioning this incident at a publishing workshop, I went into the databases to search for the authors and topic, and found that the two papers had eventually appeared in two further journals, C and D. A close check showed that they had removed some, but absolutely not all, of the most egregious overlap, and for a few days I mulled over contacting the editors of journals C and D to explain the background. In the end, I did not, but the case still bothers me.

How Not to Slice the Salami

A term used not only by food scientists is "salami-slicing," which refers to taking a study that is complete and seeking to maximize the number of publications that can be squeezed out of the data, rather than the best selection of papers, like slicing a salami into the maximum number of very thin sections. This is sometimes referred to as seeking the "minimum publishable unit," the lowest amount of content around which a paper could arguably be based, and relates to what the OECD referred to as artificially proliferating publications.

The objective of a researcher should never be to get the maximum number of publications, but rather to aim for the best set of publications, and the two are very unlikely to coincide. Excessive publication is today actually seen as a warning sign (how did they get time to do all that research?) and the problems with excessive levels of publication were well articulated in an article in 2010[21] as including: excessive demands on referees' time; mountains of often redundant reading to do on any topic; publication of minor papers; costs to libraries; consumption of resources, energy and time;[22] and forced neglect of long-standing older literature.

Sometimes, there can be very good reasons for taking a large study and splitting it into several papers, if these are substantial enough to stand alone and would not benefit from being published in some combination of the individual parts. In such cases, it is also critical to make absolutely clear that the papers are related and how, if for example they are based on the same study or set of experiments. A set of papers might be sequenced such that the first paper describes a study and set of experiments, and the later ones refer to that for the details rather than repeating. Any attempt to portray separate papers based on a single study in any way that could lead readers to believe they were from separate studies would be completely unacceptable and viewed very dimly.

Financial Fraud and Conflicts of Interest

Huge sums of money are required to fuel research in universities and research institutions, typically of the order of hundreds of millions or euros or dollars per year for a reasonably research-active organization. This money mostly flows in from external sources, like large funding entities such as national government agencies (such as the National Institutes of Health in the United States) or the European Union. Given this environment, it is perhaps surprising that, compared to other forms of research misconduct, it is relatively rare to find a case where someone has worked out a way to divert some of that into a new house or car, or just a bank account. This perhaps reflects the closeness with which such funding is typically managed and audited by institutions and funders.

However, a researcher who might not have personally profited outside their research from such funding may still have committed serious financial misconduct, for example if they have obtained such funding under false pretenses, or else used external funds to undertake research that is later found to be fraudulent. In such cases, institutions have been punished by demands for reimbursement of funding received, as when the Massachusetts Institute of Technology (MIT) was required to return $2 million to the National Institutes of Health when its staff member

---

[21] Trimble, S. W., Grody, W. W., McKelvey, B., and Gad-el-Hak, M. (2010) The glut of academic publishing: a call for a new culture. *Academic Questions*, 23(3), 276–286.

[22] A new journal in the field of environmental and sustainability science called *One Earth* offsets three tons of CO_2 for each scientific paper it publishes.

Luk Van Parijs was found in 2005 to have fabricated data in multiple papers in the area of immunology (Van Parijs was fired, underwent six months of home detention with electronic monitoring, completed 400 hours of community service, and had to pay back to MIT the amount of the grant that was already spent).

In a related consideration, researchers are expected to be open and transparent regarding conflicts of interest in their research. This relates to cases where the source of funding could have a bearing on the conduct, analysis, or reporting of a piece of research, and in all such cases the readers of a paper must be given any information in this regard, such that they can decide for themselves if such knowledge influences how they interpret the outcomes of the study.

If a pharmaceutical company paid for a trial on a drug they manufacture, and the researchers published a paper showing fantastic outcomes for that drug, the readers might interpret this differently if they knew the source of funding compared to a scenario where they thought this was a completely impartial study, where the researchers (or funder) gained no benefit from the publication of these findings.

A conflict of interest might also arise from a personal belief, such as the authors of a paper on the benefits of a particular diet not declaring that they are practitioners of that diet, or a group of authors with a particular religious persuasion not declaring that when publishing a study directly relating to that topic.

Integrity and Peer Review

The process of peer review, as described in chapter 6, is a key part of the quality control process of science, and for the reasons described there is normally performed semiblind, where the reviewers know who the authors are, but the authors do not know who the reviewers are. The possible integrity issues that can arise here relate to the opportunity for reviewers to exert unfair influence on the potential for publication of the work of other researchers, either positively (going soft on known colleagues) or negatively (harshly reviewing, or rejecting, the work of rivals).

There is also the possibility of stealing ideas through peer review, when the referee finds themselves in possession of information in advance of it appearing in the public domain. There have been a small number of reports of reviewers stealing ideas from a paper they were asked to review, rejecting that paper to slow it down, and then publishing their own paper incorporating the stolen ideas or materials. In one case, the reviewer who did that had their own paper sent after submission to the author of the paper from which they had lifted the material, leading to a very public and nasty airing of the circumstances.

Thus, while it is in my view a moral imperative that researchers contribute to the process of peer review if they expect to benefit from such activities of others, such activity is as fraught with integrity-related concerns as any other part of research. Reviewers must respect confidentiality and act as impartially as possible and, to facilitate the latter goal, it might seem likely that in future we will increasingly see double-blind reviewing where the authors' identities are shielded from the reviewers.

Is Science Fundamentally Self-Correcting?

It is often stated that one of the greatest strengths of science is that it is self-correcting, in that, when a published paper reports interesting results, others will attempt to repeat and build on the work, and if it cannot be repeated (for innocent or malevolent reasons), then it will quickly be forgotten and science will recover from the dead end down which research had been briefly pointed by the "bad" paper.

However, many studies have questioned the extent to which any paper is genuinely tested by replication in this manner,[23] leading to what has been called the replication crisis. In addition, there is a good argument to be made that attempting to repeat work that—for whatever reason—cannot be repeated is simply a waste of time, effort, and resources.

A related point here is the extent to which science can recognize and deal with fraud. One chance to do this is of course peer review, and a certain proportion of fraudulent papers might get caught before they appear in the public domain, thanks to the perceptiveness of eagle-eyed reviewers. After publication, it is up to the readers of the journals to spot and report such issues, and several problematic papers have been identified when other experts reported that the data didn't look "right."

It might be imagined that a goal for any researcher is to produce neat data sets where points are neatly on lines or curves, error bars are small and tidy, and differences between treatments very clear, but actually nature isn't so perfect, and some data sets will be deemed to look "too good to be true." There are also statistical principles that predict number frequencies in natural data sets (Benford's Law), which means that it is very hard to fake a set of data in such a way that digit distributions look genuinely natural. If anyone were asked, for example, to write 500 random numbers from 1 to 10 in sequence, it is likely impossible that they could generate a really random list without unconsciously inserting unappreciated number preferences or patterns. While scientific data are clearly not random numbers, it is likewise very difficult to invent data that reflect what nature might throw up.

This sort of reasoning has given rise to a very unfortunate branch of science which, in a pitch unlikely to be green-lit by any television executives soon, could be called "*CSI: Science Data Police,*" where individuals claim to have forensic proficiency at analyzing scientific data sets for evidence of fraud. There have been a number of cases where sometimes self-appointed guardians of scientific standards (not necessarily referees, editors, or even researchers, and sometimes as mentioned earlier so-called data thugs) have persecuted cases, rightly or wrongly, to very serious consequences based on their interpretation of data, which has led them to accuse the authors of fraud.

A key question of relevance here is the difference between honest error and intention to deceive. This leads me to one more example that illustrates many of the issues discussed in this chapter and is perhaps one of the most high-profile and damaging cases of allegation of scientific misconduct of the 20th century. The figure at the center of this scientific maelstrom is David Baltimore, a biologist who won the Nobel Prize in 1975 for his part in the discovery of the mechanism by which viruses called retroviruses (a family that includes the HIV virus that causes AIDS) insert genetic material into the DNA of cells and may thereby influence the development of cancer.

In the 1980s, Baltimore was at the Massachusetts Institute of Technology (MIT) and published a paper with a researcher named Thereza Imanishi-Kari on the immunology of mice in the journal *Cell* in 1986 (the paper is known after its first author as Weaver et al.). Following publication of that work, another researcher at MIT started to raise concerns about possible fabrication of data in that paper, based on their inability to reproduce some of the reported findings. This led to an escalating series of inquiries and investigations, particularly thanks to two early examples of "data thugs," who termed themselves fraud-busters (the original version of the movie *Ghostbusters* was very popular in that era).

The Baltimore team published corrections to the paper (claiming errors rather than any deliberate fraud), but the questions continued, going as far as congressional hearings involving politicians who became very interested in a such a high-level case of alleged scientific misconduct. The US Secret Service even became involved, and seized "evidence" including Imanishi-Kari's notebooks (which were famously untidy) to undertake forensic analysis of inks and printouts, looking for evidence of fraud.

A 1991 report by the National Institutes for Health found Imanishi-Kari guilty of fabricating and falsifying data, and criticized Baltimore for not taking sufficient actions in the face of mounting criticism. The paper was then retracted, although Imanishi-Kari and one other author did not sign the retraction, and around that time Baltimore was forced to resign his new role as president of the respected Rockefeller University, due to the negative attention he was attracting. In 1992, it was announced that criminal charges would not be pressed against Imanishi-Kari, but a 1994 report found her guilty of 19 counts of scientific misconduct, and Imanishi-Kari was banned from seeking federal funding for 10 years.

Interestingly, around this period, a number of other research groups reported that they had confirmed the results of the *Cell* paper.

An appeal was convened in 1995 to review the case and a year later essentially threw out all the charges, mainly based on problems with the Secret Service analysis (not a role they usually play, being primarily responsible for protection of the president of the United States), and also a lack of any apparent motivation for Imanishi-Kari to have done what she was accused of. The report (based on hearings that generated almost 6,500 pages of testimony from 35 witnesses) did conclude that the original paper contained several errors.

So, one paper, in one journal, turned into a decade-long investigation which derailed several careers, resulted in the resignation of a university president, brought hugely negative media and political scrutiny to bear on research, and bitterly and deeply divided the biology community in the United States into pro- and anti-Baltimore camps. Reports, articles, and online discussions on the case since can be divided into those that regard it as a case of unfair and partisan persecution, and those that believe the guilty managed to get away with it.

One paper, in one journal, and all that damage. Publishing in the modern scientific literature is not a trivial matter, and all those engaged in that activity must acknowledge the seriousness and responsibility it involves.

The Consequences of Misconduct

A key question that follows from the examples we have discussed is what the consequences and penalties for misconduct should be. At an individual level, a researcher shown to be guilty of misconduct will face consequences of varying severity, depending on the nature of the offense. The consequences may also differ depending on the stage at which the misconduct was discovered: during peer review (i.e., prepublication), shortly after publication, or a significant period of time after publication (when impact is likely to have been much greater than if it were picked up quickly).

The procedures to handle any such cases are complex and still evolving, but international guidelines have been published by the Committee on Publication Ethics (COPE), an organization founded in the late 1990s with the mission "to educate and advance knowledge in methods of safeguarding the integrity of the scholarly record for the benefit of the public." COPE have very helpful flowcharts available online with recommended step-by-step decision-tree processes for cases of FFP discovered either before or after publication.

If a paper is submitted to a journal and found (through routine screening) to have significant levels of plagiarism, the editor is likely to reject it immediately, and there may be no further consequence. However, the author's name may be then flagged within that journal, which may influence the reception subsequent submissions receive, and editors may choose to share this information with editors of other journals. So, even if public shame may be averted, the ability of authors to publish in the future may be significantly impeded.

If the issue picked up during peer-review is evidence of image or data manipulation or the presentation of data that clearly look fraudulent or suspicious, the recommended procedure from COPE is for editors, where they are satisfied that legitimate concerns have been raised by reviewers, to allow authors a chance to respond and clarify; if this does not satisfactorily resolve the matter, they should contact the author's institution, raising the matter and proposing an investigation. If this results in a determination of misconduct, the paper will be rejected, but the author's troubles closer to home are likely just beginning.

Increasingly, problems with FFP are being referred to researchers' employers, and so the likely career impact associated with misconduct is progressively deepening. There is no question that careers have been terminated or at the very least significantly negatively affected by the fact that individuals have been involved in misconduct, and research organizations and universities worldwide are taking ever more stringent positions on thoroughly persecuting such cases.

Another key outcome of misconduct being identified post-publication, as stated earlier, is the reputational damage that results from the public notice that is a retraction being readily found through database searches, as well as their naming on sites such as Retraction Watch.

The COPE guidelines for suspected fabrication or falsification in a published manuscript align closely with those for a submitted one, but obviously end in terms of the specific publication in question with retraction of the paper, and in terms of the author(s) with whatever consequences their employer may deem appropriate for the offense committed.

It has been reported that a retraction can result in a "citation penalty" for a researcher's body of work as a whole, whereby there is a measurable decrease in later citations to their work, so a retraction can also trigger broader suspicion of someone's work as a whole.

The employer of a researcher found to have committed misconduct might be punished by the funding agency that has paid for the researcher in terms of returning the (frequently substantial) funds awarded. In addition, in the United States, researchers found guilty of misconduct have been banned from seeking federal funds for research, perhaps for periods up to 10 years, a penalty that on its own can essentially end someone's career.

The most severe consequence faced by a researcher found guilty of misconduct might be a jail term. This is rare but absolutely not unheard of, with some recent examples being a US researcher sentenced to almost 5 years in jail (along with a fine of over $7 million) in 2015 for falsifying and fabricating data in a trial relating to an HIV vaccine, and a researcher in the UK who was sentenced to 3 months in prison in 2013 for tampering with data in a cancer study.

Most broadly, every instance of misconduct results in further erosion of trust in scientists by the public, public confidence in key issues (seen in cases relating to areas such as climate science and the MMR vaccine, to name but two), diminished trust of funding agencies and governments, and guilt by association of coauthors.

Finally, it should be noted that most national, international, and institutional codes of conduct and integrity for research state that concealment of misconduct of others is itself a form of misconduct. Thus, if a researcher believes or comes to know somehow that someone else in their lab, for example, is manipulating data, then they have an obligation to draw this to the attention of the responsible local authorities (such as a group head, head of department, or institutional integrity office). Obviously, such matters should never be progressed unless a very strong case can be made, and making malicious or unfounded allegations of this nature, given the seriousness of the accusation, represents yet another form of misconduct.

Are There Differences in Seriousness and Consequences?

Consider two parallel scenarios. In one, a researcher is working on a new treatment for diabetes, and publishes a paper in which they are found, after publication, to have entirely fabricated the data that led to their key conclusion that the experimental treatment had significant potential as a future treatment for this condition. In the second, a researcher publishes a paper claiming that a distant star is being orbited by a planet, but it is later found that in this case also the data leading to that claim are fabricated.

Are these cases morally equivalent in terms of the seriousness of the offense, despite both being a deliberate attempt to present information which the author(s) knew to be fake, for presumed personal advancement? In other words, does the field of research, and the potential for damaging impact beyond the integrity of the scientific literature, have a bearing on how misconduct should be viewed?

Let's take the first scenario and split it into two alternative pathways. In one, the fabrication is discovered immediately after publication, and the paper is retracted before it is widely read. In the other, however, the fabrication is not discovered until five years after publication, and in that time several articles have appeared in the popular news, raising hopes of clinicians and sufferers of a new treatment, while other researchers have spent time, money, and effort trying to build on or replicate findings for which this is impossible, resources that should have been directed toward more useful avenues of research.

In both these pathways, the author has committed the same offense (presumably with the assumption it would never be discovered) but the ripples of consequence are very different. Do these impact on how the case should be handled or the penalties that should be imposed?

When I present these hypothetical scenarios in workshops, very heated debate can follow arguing different viewpoints and cases for equivalency of impact, but I know for sure what my gut and my personal sense of morals tells me here. That is that, while every researcher in every field has to operate by a minimum set of professional standards of research integrity, the consequences of breaches of these standards can differ between disciplines, as can impact, and so the outcomes could and should be different.

How to Handle Honest Error

Of course, errors can happen in research, as in any field of human activity, and a researcher who has published a paper and discovers subsequently that they have made an error that undermines or negates their main conclusions, or indeed any other part of the paper, has one simple action to take: self-initiated retraction. In the case of a minor necessary change which is not sufficiently damaging as to warrant a full retraction, an erratum may be all that is required, but in either case it is the responsibility of the author(s) to correct the scientific record. Such an action, if properly undertaken, does not damage, and may even enhance, the reputation of the researcher(s) concerned.

One example of this which shows the absolute right thing to do concerns an English astronomer called Andrew Lyne, who in 1991 published a paper in *Nature* reporting the detection of a planet orbiting a type of star called a pulsar, which at the time was the first report of a planet outside our solar system. This was, for that reason, a huge discovery, and Lyne was invited to speak at a major conference of the American Astronomical Association to discuss the discovery. In preparation for his talk, he realized that one possible explanation (the effect of the earth's motion around the sun) had not been taken into account and, when this was, the evidence leading to the claim for a planet disappeared.

Try and imagine the feeling that must have ensued when he discovered that, not only had he published a widely publicized (cover article!) paper on this discovery, but he was going to stand up in front of his peers as the star (ahem) of a major conference and talk about it! What he did next is held up as a case study in exactly how to properly and with dignity handle such a catastrophic (non) discovery, which was that he retracted his work publicly at the conference (uttering the words "Our embarrassment is unbounded and we are very sorry"), which received a standing ovation, and the next day sent a letter to *Nature* entitled "No planet orbiting . . . ."

Conclusions, Solutions, and Progress

There has never been as much awareness of the nature and significance of scientific misconduct as there is today. Safeguards have been implemented worldwide, in terms of codes of conduct, requirements for training, especially of young researchers,[24] and recognition of the roles and responsibilities of mentors and supervisors, as well as practices such as postpublication peer review and the roles of sites (such as Retraction Watch) and individuals in bringing cases of clearly unacceptable conduct to light. Many universities and institutions today have offices or officers

---

[24] In many universities, all PhD students must undergo training in research integrity, and progression through the degree may be contingent on certified successful completion of such training.

responsible for overseeing this area, undertaking investigations where necessary, and providing advice and input in response to queries.

How the system can best work, in my view, is that we must assume honesty and punish betrayal of that trust brutally. Imagine a world in which accepting an invitation to review a paper resulted in a knock on the door to find someone bearing a box of printouts, handwritten notes, USB sticks of data and other evidence of raw data that a reviewer was required to work through in order to validate that every point in a table or figure represented real and verifiable data. Would that work? No, unless the reviewer really wanted to work from such first principles, so we must assume (unless there is very clear evidence to the contrary) that submitted papers are real and honest, extend the benefit of the doubt and presumption of innocence to the authors, and then punish abuses of such trust with the full weight of consequences discussed earlier.

On this rather glum note, let us turn back to the practices of the honest and professional researcher in communicating the outcomes of their research, and turn from publication to presentation.

{ 8 }

Conferences and Presentations

Conferences: The Social Glue of Scientific Fields

While the gold standard for communication of new scientific information is the article published in a peer-reviewed journal, there are several other ways in which scientists communicate their findings.

The first of those is through conferences, where they present short accounts of their work, typically in advance of publication of the relevant journal article, in either a "poster" or oral form (a presentation lasting anywhere from 10 to 30 minutes for a research talk, or longer for a talk, typically invited, in which a researcher gives an overview of an area in the form of a review presentation).

WHY DO SCIENTISTS PRESENT AT CONFERENCES?

First, the discussion and questions that follow a presentation allow the kind of immediate and real-time feedback and discussion that is more difficult for written forms of communication (although preprints, blogs, and other online fora are increasingly providing other routes, albeit more indirect, for doing this).

Second, presentation at conferences can present an opportunity to stake a competitive claim on an area, indicating progress and demonstrating results much faster (albeit more ephemerally) than in a journal article.

Finally, and crucially, conferences (or smaller workshops) are, in every field, part of the social glue that holds a community of like-minded researchers together, offering a chance for those working in a field, from the newest students to the most established names, to come together for a few days to meet, network, discuss, collaborate, argue, and make or renew acquaintance (as seen very nicely in chapter 6 for the story behind the discovery of buckminsterfullerene).

How to Find the Right Conference

As with the proliferation of journals of various repute in recent years, there has been a dramatic increase in the number of conferences being organized, and established researchers are bombarded almost daily with invitations to attend or present at events around the world.

Unfortunately, as with the increase in predatory journals which could entrap the unsuspecting researcher, many of these new conferences are likewise not to be considered. Like many researchers, I am bombarded with invitations to speak or present my work at conferences around the world that are highly questionable.[1] In some cases, researchers have attended such conferences and found them to exist, but typically to be poorly run and with programs not resembling those advertised.[2]

Some clues a researcher should use to short-list conferences in terms of considering attending or not would be as follows:

1. Organizers: Who is the invitation from? Is the conference being organized under the auspices of a professional society or learned organization, or is a major publisher backing and organizing it (as happens increasingly frequently)?
2. Announced speakers: Are the speakers announced to date as being confirmed of high standing within the field, and interesting to hear?
3. Objective and cost/benefit analysis: What is the opportunity offered by the conference in question? Will the researcher have an invitation to present (the best option, as this might include financial support for travel, accommodation, registration, or all of the above) or submit an abstract as a request to present a paper or poster? Will the likely benefit to the researcher in terms of networking, contacts made, and opportunity to highlight their research justify the cost incurred (both financial in attending but also in time spent preparing and attending)?

Preparing a Scientific Presentation: Planning around the Audience

As for all forms of communication discussed in this book, the key initial question on planning any scientific presentation is simple: Who is the audience?

To unpick this a little, we could ask:

(A) Who is the audience?
(B) What do they know already about what I am talking about?

---

[1] I once received an invitation to speak on 3D printing of cheese at a conference on regenerative medicine and stem cells. I was also recently invited to a conference in Rome on the basis that my valuable insights could be very instrumental in motivating and inspiring the conference attendees to, among other benefits, "ignite them to become the torch bearers of the society." Flattery may get you everywhere, but it didn't get me to Rome.

[2] An account of such a conference attendance can be read at: https://www.technologynetworks.com/tn/articles/inside-a-fake-conference-a-journey-into-predatory-science-321619.

(C) What do I want them to know?

(D) How can I most effectively get them to (C), taking into account (B)?

The same researcher could hypothetically present on the same topic to multiple different audiences on consecutive days, from the most expert to the least, from peers to the public. Knowing how to judge the level at which to pitch a scientific presentation to different audiences is a key skill for a successful researcher, and in the next chapter we will focus on presenting to nonexpert audiences. For now, let's assume a level of technical expertise.

Even assuming the audience are all relatively specialized, however, doesn't necessarily simplify things completely. For example, are they all of the same discipline as the researcher? Is it a small audience all of whom are familiar with the type and context of research being presented, or a larger mixed audience where there are several cohorts present who might each be familiar with all, some, or none of the material?

A key concept in considering a presentation in this regard is that of the remote control, and specifically its absence in a live talk. When reading a paper or book, a researcher can pause to check something in another paper or by googling, rewind if they need to check something earlier in the same piece of text, or fast forward if they want to skip over something they don't need to know, know already, cannot understand, or find boring or irrelevant.

In a live presentation, however, the audience are essentially captive, with none of these tools at their disposal, and so the presentation must unfold at a pace that means the lack of such interventions is not a problem.

A presenter fails if their audience (or subsections thereof) become lost, bored, or confused, and judging the audience carefully in advance is the best means to avoid such undesirable circumstances from arising. Thus, the speaker must judge the exact backgrounds of the audience and subcohorts thereof, and do their best to neither bore those who are closest to the material nor lose those who are furthest away, as it were.

This is not always an easy needle to thread, but needs to be a driver for starting to approach the presentation, particularly in the first sections. For a homogeneously expert audience (if such a thing exists), it could be safe to jump straight into detail and jargon, and assume the audience know the relevant literature, acronyms, and so forth. This would be a disaster, though, for those who are coming relatively unprepared to the talk.

Preparing a Scientific Presentation: Building a Skeleton Talk

Having identified and analyzed the audience as much as possible, the next step might be to prepare a skeleton of a talk, literally starting with a wire-frame outline of slides and indicative content. This leads to the immediate question of how many slides might be needed for a talk of a certain duration.

A remarkably useful rule-of-thumb is to aim for one slide per minute allocated, and it is amazing how often this turns out to be exactly the right number of slides to fill that time—fifteen minutes, fifteen slides, twenty for twenty, and so forth, giving let's say "n" slides to work with.

Let's see how that might break down for a research-focused (as opposed to a general review-type) talk. One slide will be title, and one conclusions, with perhaps one at the end for acknowledgments, logos, and so forth. That leaves n – 3 slides to fill with content.

At least one or two will need to be introduction, depending largely on how much background is needed to bring the audience up to speed (based on how familiar they are likely to be with the material, as discussed earlier).

Then, in terms of results to be shown, the evidence or artifacts could be laid out that "must" go in, perhaps a couple of tables (to be formatted or constructed afresh), some images, and plots. This then reduces the number of slides accordingly, and eventually it will become very clear how many slides or parts thereof are left for things like accompanying text, experimental design, and so on.

A key consideration here is the overall length of the talk, which will determine exactly how it might be further substructured and divided (or not) into sections. For a short talk, a useful concept is that of "zooming," where the start of the talk should be general and broad, and then the presenter gradually "zooms" into greater levels of detail as the specifics of the work undertaken and results obtained are discussed, then "zooms out" at the end to consider the big picture again, in terms of conclusions and implications.

For a short talk, maybe 15 minutes maximum, an audience is likely to be able to spend around 80% of that time concentrating on details and handling plots and tables in rapid, one-per-minute, succession.

However, if the length of the talk goes much beyond that, it is unlikely that even the most committed audience will all manage to maintain that level of concentration. Thus, if a talk is 20, 30, or 40 minutes, and contains a significant amount of detail in that duration, the speaker will need to break up the flow deliberately with "zoom-out" pauses, perhaps every 10 minutes or so, where they reduce the level of detail and intensity of the topic to allow the audience to breathe a little easier due to having to concentrate less intensively. This could be achieved, for example, by inserting an intermediate conclusion slide, summarizing the material discussed in the preceding section, or possibly by including at the start a slide with a map of the flow of the talk and sections thereof, and then putting back up that slide at the required "break points," ostensibly to indicate what the topic will be, but also allowing a respite, however brief, from the onslaught of detailed information.

These considerations will rapidly build up the internal architecture of the talk, and create a skeleton of content and slide-by-slide indicative content.

A key next step here might be to review what material is already available for incorporation, or what needs to be generated de novo. While a key principle in scientific writing is the avoidance of recycling text, in the case of presentations it would be a very difficult situation for researchers if there were a similar principle that any slide, once used in public, could never be reused. Of course, when speaking on a specific topic for the first time, a researcher needs to generate all slides from scratch, but careful time spent on their design and creation will usually be repaid by the potential to reuse these in similar or modified form in later presentations.

A reason why this might be acceptable for slides in a way that it is not for text is that every written research publication is a permanent contribution that remains available into the future, and so there is no justification for having the same text available, broadly speaking equally, in different documents. A scientific presentation, by contrast, is inherently ephemeral, and seen only by a specific audience at a specific time, which is unlikely to be the same the next time those slides are displayed. Of course, if someone gave the same talk repeatedly to the same audience the repetition would be obvious and likely very annoyingly boring and inappropriate, but this is unlikely to be the case. Thus, as researchers gain experience of presenting, they also build up a master slide deck of material which they can dip into to modify or extract from as needed for future talks.

A THREE-DIMENSIONAL CASE STUDY OF CHEESE

In February 2017, my research group published a paper on the impact of 3D printing on the structure and texture of cheese. This arose from a series of studies inspired by a query from a food company about the long-term potential for use of cheese as a raw material for these futuristic devices (perhaps to be found in domestic or restaurant kitchens, as well as industrial production environments) about which excitement was gradually building. We were happy with the study and the paper, and I submitted it, one of perhaps 20 others that year, from which it stood out for little other than its quirkiness and the fun we had (with a series of visiting French students) doing the experiments.

I was not prepared for what came next, which was a huge level of interest on social media (see chapter 9 for more details on that aspect). While all this was an entertaining new experience for me and the others involved, and a source of amusement for friends and colleagues, the work attracted more serious and professional interest, including a number of invitations to speak about the work at international conferences (some of which I was already attending or speaking at), reflecting clear relevance and timeliness of the work (or perhaps an occasional need to have a lighter talk among more serious fare).

The reason this example is relevant here is that, while the topic for all the talks was the same, the audience was different in each case, and I had to design each presentation differently based on the audience, as summarized in Table 8.1. If I represented the five talks in that table as a Venn diagram, they had a big overlap

TABLE 8.1 Five Dimensions of a Presentation on Three-Dimensional Cheese

Conference	Location	Audience and Impact on Presentation
A, June 2017 (Workshop)	The Netherlands	Workshop was dedicated to 3D printing of food, so I could assume a good level of 3D printing knowledge (including from earlier presentations in the workshop), but little or no expertise in cheese or dairy products. Audience was a mix of industry and researchers (around 100). Could skip 3D background but extended introduction to cheese science and gave moderate level of scientific detail.
B, September 2017 (Trade fair)	Germany	Audience (around 50) was mainly industry, with little awareness of 3D printing, some of dairy science. Expanded 3D background and introduction to cheese science, moderate level of detail throughout.
C, October 2017 (Conference)	The Netherlands	Mainly researchers, little awareness of 3D printing, high level of expertise in dairy science (around 150). Expanded 3D background, skipped introduction to cheese science, kept high level of detail throughout.
D, November 2017 (Conference)	Northern Ireland	Experts (around 200) in dairy science, mix of industry and researchers, little awareness of 3D printing. Expanded 3D background and introduction to cheese science, kept high level of detail throughout.
E, March 2018 (Conference)	Ireland	Experts in engineering and technology, less so on 3D printing, mix of backgrounds and levels of expertise on dairy science (around 120). Expanded 3D background and introduction to cheese science, moderate level of detail throughout.

in the middle, and perhaps 40%–50% of content was identical, but the introduction and final conclusions for each differed hugely based on (1) what the audience was expert in (and how expert overall, and able to handle technical details of methodologies or printers, they were), (2) what the audience was interested in, (3) who they were (academic, industry, engineers, food scientists, and so forth), and (4) how long I had to tell my story (between 20 and 30 minutes). There is no way the same talk would have worked for all groups, when some knew all about cheese and nothing about 3D printing, and some started in exactly the opposite position.

This is as clear an example as I could give of tailoring each presentation based on a very careful analysis of the target audience, and I won't pause here to give the many examples I could cite of hearing speakers who failed to do this and delivered talks that were completely misjudged for the audience in question, either soaring over their heads with a level of specialized detail they could not handle or boring them by lingering over very basic information.

The first rule of every presentation must thus be this: consider the audience.

DESIGNING THE SLIDES

In the sections that follow, I am going to assume that presenters are using the dominant software type used for preparation and delivery of presentations, which is Microsoft PowerPoint,[3] but note that other presentation software is available (such as Apple's Keynote, which is broadly similar to PowerPoint, and also less common but creative packages like Prezi, where text and images can appear in manners very far from the linear transitions of PowerPoint).

The next step, when the initial talk skeleton or storyboard is in place, is to start to design the visual look of the talk (by which I mean background, color scheme, font and font size, and so forth). This might be influenced in part by the use of certain slides that have a particular design, which might make it easiest to work with that style. Some researchers might also have a preferred style they use, with which they wish to be consistent, or (more likely) may be working within a context in which they have been provided with a format to work within.

For an example of the latter, many universities or research centers may have (sometimes professionally designed) slide templates that all their staff or students should use, displaying the organization's name and logo prominently (and hopefully not swallowing too much of the available slide space in the process). I have occasionally also encountered conferences that have standardized slide designs they wish speakers to use, sometimes only for the title slides (to brand the talks as being of that conference) but sometimes also with less complex designs for each slide after.

If such constraints are in place, this needs to be known and worked around from the very start, as retrofitting slides onto a new template and moving text, images, and suchlike around to get out of the way of colored zones or logos is intensely annoying, and is close to having to start again in terms of slide design.

If the presenter has effectively a blank canvas to work with, the question of overall slide color scheme or background design must be tackled. It is believed that dark backgrounds with bright text are better in small rooms while pale backgrounds with dark text work better in large rooms, but there are a number of other factors that can affect the visibility and clarity of information on a screen.

For instance, ambient light, windows, blinds, curtains, lighting fixtures near a screen, and many other environmental factors can influence how clearly slides appear on screen, as can something as simple as the strength (or lack thereof) of the bulb in a projector that is beaming the presentation. Slides that look extremely clear on a computer screen, in particular in terms of colors and color contrast, can

---

[3] This tradename as written is an example of Camel case, where the text mixes upper and lower case letters in a single name, like the humps on a camel. It has been quipped that PowerPoint is so called because so many presentations using it contained neither (thanks to John Finn!)

look very different when projected 10 feet high on a screen in a bright room with a projector the bulb of which is many hours past its optimal replacement time.

What this means is that, where possible (and it may not be), someone giving a presentation really needs to scout out the location in advance, and ideally see and practice their talk using the exact equipment and space where the real thing will happen. Walking into the room where you are to give a presentation minutes before is not to be recommended, if circumstances allow this to be avoided.

It is all about minimizing what can go wrong. One of the most likely reasons for anyone to be nervous is fear of messing up, and so every precaution should be taken to minimize the risk of this happening, and reduce the unknown variables. Reconnaissance of the location is one part of this, but it can go further, for example into control of the audiovisual options to ensure that they will work with complete confidence, by a researcher bringing their own laptop to connect to the projector (probably not allowed in large conferences but quite likely to be allowable in more informal or seminar scenarios) and having their own remote control for controlling the PowerPoint slides, which they know absolutely how to work without hesitation, and for which they even have their own spare batteries. Using the researcher's own laptop also allows them to have the presentations set up to give them the most useful real-time information, such as the laptop monitor being set to show not just the current slide but also a split screen with the current slide, the next slide, the time since the presentation started, speaker notes, et cetera.

That is not to say that nerves are bad—I will always be nervous before any presentation, and would be actually worried if I wasn't! A little nervous tension is nothing to be ashamed of, as it focuses the mind and adds a certain adrenaline-fueled tension to the situation, which hopefully leads to greater alertness and fixation, bringing the presenter fully into the moment with only one thing in mind, and a burning desire to do that thing as well as they possibly can.[4] I can think of occasions where I have given presentations on days where there is other stuff, concerns or worries, on my mind, and only realize how effectively they have been pushed out of my mind in the minutes before and of the presentation when I feel the almost visceral sensation of awakening as they flood back some point later when the single-minded focus starts to dissipate.

Of course, the key is to have sufficient nervous energy to achieve this focus without it being so severe that making the presentation becomes a hugely stressful or frightening experience. When young researchers ask how to avoid this, the best pieces of advice I can offer, as this is a hugely personal phenomenon that everyone reacts to and handles in a different way, are, first, minimize the potential for fear-inducing crisis by the type of preparation steps mentioned already (and of which more later), and, second, practice as much as possible, not only in terms of the talk

---

[4] There is apparently a trick used by some British politicians to increase the intensity and urgency of a key speech or meeting; I won't go into details here, other than to note that it is named the "full bladder technique" (not to be confused with the term P-hacking mentioned in chapter 3).

in question but also in general, in terms of standing up and presenting in front of an audience.

THE FOCUS OF A PRESENTATION AND
THE VEHICLES OF COMMUNICATION

When planning a presentation, a key question is what is the intended persistent "take home" message. What would a presenter expect an audience member who has paid attention to recall one day after the presentation, one week, or one month? If this can be identified, this becomes the target around which the rest of the talk is structured, like a road being built to reach a specific identified destination.

How does a presenter know if they are getting there successfully, or if there is too much potential for the audience to get lost, side-tracked, or simply give up? One good tip I came across for evaluating this is to practice a talk in front of an audience of colleagues or fellow students in advance and omit the final conclusion slide or equivalent. Then, once the rest of the talk is finished, the audience are asked to write down what they think would have appeared on that slide. If they match what the presenter had intended to put down, then all is good but, if they do not, then something needs to be fixed.

A related question to this is what the main vehicle or medium of communication of that information is. In a PowerPoint-based talk, there are two separate media by which information is being transmitted, which are the slides and the presenter's voice. Which is the primary medium of communication?

In my view, this is simple; it is the presenter's voice. This was first clearly brought home to me when I attended a conference on the history of science at which some talks were presented by historians and some by scientists. The latter gave a familiar (to me) style of talk based on slides with accompanying explanation and discussion, while the former sat down and read out what seemed to be effectively very well-prepared and articulate speeches. This difference between presentation modes in different disciplines, I came to understand, is quite common (although perhaps gradually changing as PowerPoint, for better or for worse, pervades more and more widely), but it emphasized to me that talks are called talks for a reason. It would be pretty weird if someone did the opposite and stood silently while simply advancing slides on a screen.

If we think about what a presentation should do, it should educate, explain, argue, convince, enthuse, and perhaps entertain. Can words or images on a slide do this? Perhaps, probably not, but certainly not as well as a good human can.

To me, what appears on slides is simply the evidence that cannot easily be spoken but must be shown (figures, images, tables, data) plus prompts to remind the presenter what to say at each stage of the presentation. The presenter must present and control the material, while the slides support but do not replace. This even comes down to sequencing the rate of appearance of information on individual slides—if a text-heavy slide is unavoidable, then putting all text up at once will result in the audience switching their attention from the speaker to the slides, and not listening but reading. The presenter in such cases can best maintain control by controlling the rate at which text appears, perhaps line by line, so that they are the focus, and the audience only reads on the screen what and when they want them to.

EFFECTS AND ANIMATION

It is perhaps hard to believe but the modern use of PowerPoint, as with many aspects of scientific communication discussed in this book, is a relatively recent phenomenon, and became widespread only in the late 1990s. When I was a PhD student earlier in that decade, preparing "slides" meant printing each of those onto small (maybe one inch by one inch) translucent acetate or celluloid rectangles mounted in cardboard frames, which were then loaded in sequence onto a circular carousel. This was in turn mounted onto a projector, which beamed light through each in turn, at a manually controlled rate, onto the presentation screen.

When the idea of bringing a computer (floppy!) disk with preprepared slides to load onto a computer was first an option, this was such a huge advance that many presenters got very carried away. Early scientific presentations were dizzily head-spinning confections of crazy colors, highly creative animations, zooming and flying text, and often cinematic sound effects. Eventually, calmness descended and it was accepted that, for a typical effective scientific presentation, we probably need to use 10% of the functionality offered to us by PowerPoint.

The animation potential of that software remains very effective in controlling the release of information, and the rate thereof, which is a key part of a good presenter's skill. Text, as mentioned earlier, can be drip-fed at the correct rate to maintain focus by animation, and so can diagrams and schematic information, which ultimately becomes quite complex through being gradually constructed by stepwise introduction of elements or layers of detail. In addition, animation can be used to guide a viewer through complex figures or tables, by highlighting critical elements which the speaker is describing using color, surrounding boxes, or shapes, or even simple "look here" arrows.

A FONDNESS FOR FONTS

While there needs to be a rational minimum of text on slides (as the sentences containing the explanatory detail are verbally, not visually, presented), there must be some, and this requires a font (but definitely not several, changing between slides or sections) to be selected.

I think that some years ago there was a fondness for using Comic Sans, but it gradually seemed to be accepted that this was excessively informal or frivolous for serious scientific use. The severity of this fall from grace was nicely illustrated in July 2012, when a senior Italian scientist at The European Organisation for Nuclear Research (CERN) made one of the key presentations at the announcement of the Higgs Boson using Comic Sans; this led to a somewhat disproportionate backlash from a lot of the physics community who felt that the momentousness of this epochal announcement was somewhat spoiled by that damn font.[5] CERN did take the feedback to heart, though, and the following April (the first day of the month, in fact) temporarily relaunched their website in that font.

Another font widely used in normal scientific (and other, such as newspapers and books) writing is Times New Roman, but it is also not recommended to use this font for presentations, because the font is seen as excessively fussy and ornate,[6] and so simple plain fonts like Arial, Calibri, or Tahoma are probably most preferred. These are called "sans-serif" fonts, as serifs, the extra ornamentations found in the more complex fonts, have been accepted to work better for reading on a page than for the rapid assimilation of information in a presentation.

Letters in the optimized fonts take up less space, and are less demanding on the eye (supposedly) such that fewer neurons are required to fire and fewer microexpressions of concentration are required on the faces of the listeners, who are supposed to spend as little brain power as possible on reading during a presentation, when they should be devoting that to listening and thinking.

Another type of text that is generally to be avoided or minimized in presentations is CAPS TEXT. One reason for this is the perception that use of such text is the equivalent of shouting at the audience, but a more subtle factor is that apparently, when we read text, our brains do not actually digest and identify a word by reading each letter in sequence, as might be expected. Rather, we take short-cuts by identifying word shapes and forms, and know the shape of common words because we know that some letters have bits that stick up, while some have bits that curl downward, and together these give familiar shapes that label a word before a letter is read. On this basis, the extremely powerful supercomputers that are our brains skip through text by familiarity with the easy words and only have to pause and think harder about the more complicated or less familiar ones. Putting text into caps, however, deprives our brains of these shape cues, and extra fractions of seconds are required to analyze and "read" words, and so more of the viewer's attention is sucked into that slide that would otherwise be free to listen and think. Caps text simply swamps our processing bandwidth.

SELECTION OF IMAGES

If we assume that a typical scientific presentation will involve more than text, then that which is not text might be tables, figures, or images (such as photos, schematics, or maps). There will be two types of images that are included on slides: (1) those which are there as critical pieces of evidence (an image that explains the research design, or shows a result, for example a medical outcome) and (2) those

---

[5] An online search for articles from around this time leads to presumably (hopefully) tongue-in-cheek headlines like "CERN Scientists Inexplicably Present Higgs Boson Findings in Comic Sans," and an online petition to rename the font "Comic CERNs."

[6] Just look at the pointy bits on those letter "u"s and the aerodynamic-looking slopes and pointy bits found on simple letters like "t."

that are not mission-critical data artifacts but rather are there for illustrative and creative reasons.

I have no problem with aesthetic decisions to brighten up otherwise text-only slides with an image which is not content-critical, which might be a map, picture of a piece of equipment used, coauthor photos on first or last slide, a picture of an animal or plant in a zoology- or botany-related talk, or a beautiful-looking picture of a photogenic cow or piece of cheese in a talk on dairy science. In such cases, however, some simple questions must be considered, as follows:

- Will the image be visible? If searching online for images, always preface the search term with "hi-res" to ensure the best image quality where available, and then make sure the image when placed on the slide is not reduced so much in size that it becomes annoyingly indistinct. If that happens, instead of listening to the presenter, the audience at that point will be squinting at the top corner wondering "what the hell is that?" (it happens!);
- Will the image fit the background? Against any colored background, whether white or black or otherwise, it can work well to search for images with that background mentioned in the search (for example, search "high res image cow white background") so that a frame can be removed and the pasted-in object appear to integrate much more nicely into the slide design;
- Is the image free to use? Can we casually browse the Web and grab any image we like the look of to paste into our presentations without consideration of attribution of source or copyright concerns? It is important to be mindful of such considerations and search specifically with them in mind;
- Is the image appropriate for the audience? If the audience is professional and expert, cartoon images may be deemed unsuitable, but for a lay audience the opposite might apply. I also once saw someone give a presentation in a competition focused on nonspecialist audiences who included an image related to veterinary medicine which, while highly relevant to the topic in question and which probably would have elicited no reaction from a professional audience, resulted in audible gasps and visible recoil in their seats from the more general public audience on that occasion. They didn't win the competition!

To Practice or Not to Practice

When the slides are prepared, and a presenter is reasonably happy with that draft, it is critical to do at least one practice presentation. This will ensure first of all that the time limit specified is not being exceeded (the first rule of presentation etiquette) and, equally importantly, that the presentation works and flows. What

I mean by this is that such a run-through will make clear whether transitions work, both technically in terms of animations and slide changeovers and, more importantly, logically, in terms of being able to smoothly move between discussing slide 8 to discussing slide 9 without pausing or saying "eh . . . ."

I have never done a practice run of a presentation that has not resulted in some changes to slide order, animation, or wording (typos and errors become even more obvious somehow when on the screen), or deletion or creation of slides. This can all be done by the presenter on their own (and in my case is often in a hotel room the evening before the talk is to be given) but presenting in front of a friendly audience (lab-mates, supervisor, or students) will allow critical constructive feedback from the listener's seats, and also should include an invitation to think of questions that might be asked based on what has been seen and heard.

A key question then is whether one practice run is enough, or if more are better. In my view, if the presentation changes more than modestly after the first run, a second will be needed to reassure the presenter that the time is still (or now) met and that the changed talk flows equally well or better. It will also add confidence of knowing what to say at each point, and make sure the prompts and material on slides correctly triggers the correct commentary.

If the material is very unfamiliar, as in the first time I have presented on a particular topic or piece of research, I might do it more than once or twice to reassure myself and build my confidence. In recent years, I presented work on a study my group conducted on human milk to a pediatric science conference in the United States, and I broke my own record by practicing it four times (twice in front of different small audiences) because:

- This was the first airing of results from a major study;
- The presentation was in front of an audience whose background was very different from mine (many of whom were doctors, whom I find naturally intimidating);
- The time was very tight (13 minutes);
- I had applied to speak, rather than being invited to speak.

A key question then is the extent to which practice should continue to the point of the speaker being almost word-perfect in what they are going to say. I believe there is a risk in going too far in this regard, as the presentation delivery might then come across as "learned" and not spontaneous or natural, and I think an audience will pick up on this.

Having said that, though, there are two parts of a presentation where I would willfully divert from this principle, and these are the first and last minute of the presentation. Just as the best and second best place to be in an author list on a paper are first and last, respectively, the most important parts of any presentation are the first minute (where you try and engage and win over the audience) and the last (where you leave them with your final key message and lead into the transition to questions). So, I practice the wording for these two minutes over and over in my head, and can often be found immediately before a talk pacing nervously as these swirl in my head, and are twisted and knocked into the shape with which I am most happy. If I can then nail the opening and feel I have done what I planned, and hopefully elicited any expected reaction, then I can relax into the rest of the talk and be led through it by the prompts I placed on the slides to follow. Then, just like a pilot managing a smooth take-off, the level of timing and preparation need to be such that, perfectly on time, you can stick a perfect landing by being able to wrap up the talk in a conclusive and effective manner.

There are simple tools and tips to keep timing on track also. One presentation remote control I bought (and then sadly lost) had a function whereby you could set a time "alarm" such that the controller buzzed in your hand after a set time, so you could use it to tell you that 18 out of your target 20 minutes were gone and you needed to be preparing for your grand finale. Today, I do the same using my smartwatch (or phone set to vibrate silently in a pocket after a preset time), and such steps can keep you confidently on track without a need to distractingly glance at a room clock (if present) or watch.

As one final comment on the topic of preparation, one thing above all else that will make someone a better presenter, and that is both easy and remarkably difficult to do, is to watch video of their own presentations. It is trivial today to set up a phone to record video, but when a presenter sits down later to watch this, they experience an out-of-body experience where they see someone else who looks a bit like them, but hesitates far more than they do, faces the wrong way, scratches their head, waves their arms, or does a million other things we don't realize we do. Such a perspective can completely change our practice, and inevitably for the better. When I recommend this to students, they inevitably absolutely dread the idea, but of course if you expect others to watch and listen to you then of course it is only fair to be willing to do so yourself.

What Does the Audience Expect?

To think more about what makes a good scientific presentation, it is worth considering what a person listening to such a presentation might regard as the hallmarks of success (what does success look, and sound, like?). I would suggest four things are key here:

1. Enthusiasm for topic: it is very hard for an audience to generate enthusiasm for a topic on which the presenter seems to be themselves indifferent, and so it is critical for a speaker to be enthusiastic about and interested in their work. This is key to generating empathetic interest in the audience and capturing their attention;
2. Content: the first thing an audience member wants from a talk is new information which is of interest to them;
3. Clarity of presentation: having come for a talk on something they want to learn more about, the audience expects that to be presented in a manner

they can understand, which includes both the clarity and diction of the speaker (no mumbling!) and the legibility, design, font size, image size, and similar properties of their slides;

4. Confidence and competence: the audience needs to feel that the speaker knows what they are talking about. The more they hesitate, pause, or look at the screen rather than the audience, the more the audience will sense a lack of familiarity and confidence, which undermines the belief they will have in the speaker's message. Confidence alone, however, is not in itself enough to win over an audience, as we can picture politicians or their spokespeople who stand up and talk with aggressive levels of confidence but yet fail to convince some or all of the audience of what they are saying, and their words may be rejected as fast as they spill out.

Confidence works only where it is combined with credibility, and does not trump it. The audience must believe in the integrity of the speaker and feel that they can be trusted. Of course, achieving this balance is not something about which there is an easy prescription. Simple things can chip away at it, though, for example if a speaker makes a mess of questions, or apologizes in their presentation for anything. A commonly cited principle (which everyone still breaks from time to time) is that a presentation should never include anything for which a speaker needs to apologize (such as the classic line "I know you can't read this slide/see this figure but . . . ") as this doesn't engender sympathy as much as undermine confidence because the speaker appears not to have prepared such that, as might be expected, they don't have anything to apologize for.

A talk could be marked on each of these four-dimensional axes, and found, for example, to be a very good topic poorly presented by a deathly dull speaker, or a great speaker with an uninteresting message and who seemed to not be on top of their topic; and so the objective has to be to score well on all four criteria.

Final Steps of Preparation

As mentioned earlier, it is never a good idea to enter a room in which a researcher has to give a presentation just before they are due to do so, and so where possible getting to the venue in plenty of time to see it (and in an ideal world check slides or do a run-through there) is hugely advisable. Such a check should also examine amplification options—knowing that you need to stand still behind a podium because that's where the microphone is requires a different mindset to knowing that you can roam freely, remote in hand, because the room is small enough or else because you will have a portable microphone attached to you.

Another thing to check is whether a talk is being translated in real time, as may happen for conferences in a different country but mainly for a local audience. Think for a minute about what is required to translate in this way; the translator must listen to complex information (about which they are almost certainly not an expert), change its language in their head, and then speak it in the new language, while at the same time listening to the next sentence and doing the same, so that several processing operations are ongoing at the same time. This is to me an almost impossibly difficult thing to contemplate doing in a coordinated manner, like rubbing your tummy and patting your head at the same time (try it!). If someone has work with a translator, they should take basic courtesies like meeting the translator beforehand and showing them the slides so they have an idea what to expect, and then simply going slower than the researcher might otherwise do (in some such cases, I have stuck post-its to the monitor screen saying, in big letters, SLOW DOWN).

There are of course many horror stories of translation going awry because a speaker went too fast, as when I heard some French speakers at a major international conference in Paris speak so quickly that the poor translators were reduced to finishing every second sentence with the very scientific phrase "et cetera" while they struggled to catch up. A former colleague of mine once told of occasional laughter in a Japanese audience as they listened to a translation of their talk on polysaccharide structures; apparently important references to "double helix" as a structural motif were being lost in translation as "twin helicopters" (not the title of a classic book, at least on science).

A final consideration is dress code, and what is the appropriate mode of personal comportment for a given occasion. The key here is to judge the level of formality or seriousness expected and, if in doubt, err on the side of more business casual rather than more casual. For example, I believe that it is better for me to be the only one in a tie rather than be the only one not. I would also add, from personal experience, that one should never travel to a conference or other presentation occasion in clothes in which they would not be prepared to give the talk if the worst-case scenario (lost luggage) came to pass. Trust me on this one!

The Risks and Benefits of Humor

There is no doubt that one of the best ways in which to get an audience to empathize with and like a speaker is to use mild and well-placed humor, but this can be a risky strategy which, if badly executed, can result in disaster.

A presenter could include a carefully planned and perfectly executed masterpiece of scientific humor and find that afterward the main thing audience members remembered was that joke. At the other end of the scale, a poorly executed joke can result in the demolition of a presenter's aspirations to being taken seriously as a credible speaker, and so either extreme of success in the deployment of humor has its risks.

I once gave a talk on getting postgraduate study off to a good start to an audience of around 150 new PhD and Masters students, in the 5th minute of which I showed a slide with a visual joke on it which, when I looked into the audience,

I realized was potentially quite insulting for at least one person present. While the other 149 individuals probably didn't notice, I spent the remaining 35 minutes wishing I could crawl under the podium on which I stood and hide, and photos of the talk, in my view, show a deeply uncomfortable speaker. Be careful when taking risks!!!

On the other hand, I once had my luggage lost and ended up giving a talk at an Italian conference in an ensemble on the casual side of comfort, but began by explaining that my luggage was lost. In an inversion of the strange advice sometimes encountered for speakers to render their audience less intimidating by imagining them naked, I asked my audience to imagine me wearing a nice suit. They laughed (with me, hopefully, if the translators caught my meaning, and not at me) and I had defused the moment and could move on less self-consciously.

There is also the possibility to aim for maximum nerd-dom to win over the audience with empathy for humor that could only possibly be regarded as effective within narrow disciplinary boundaries, as when I countered someone else's much-commented-on t-shirt at a conference which said "make cheese, not war" with a line on a slide which read "all we are saying is give cheese a chance" (a dairy conference being the only place where one can get away with such cheesy humor).

What Can Go Wrong?

When PowerPoint first became a commonplace feature of scientific conferences, it was notoriously unreliable, as was the hardware, and presenters frequently had backup options of acetate slides or other far more primitive forms of displaying their material. In time, such contingency plans became less necessary, but problems can still occur, and part of preparing for a presentation should include imagining possible pitfalls and preparing plans for how to deal with these.

Some possible issues might be as follows:

1. *Presenter meltdown*: this might result from nerves or lack of preparation, and manifest itself in extra-long pauses or failure to exude confidence (for example, failing to make eye contact with the audience). The only remedy for this, partial as it may be, is practice, both at a macro level in terms of taking every opportunity to stand up and speak about their work, to build broad experience and confidence around giving talks, and at a micro level in terms of rehearsing the specific talk in question enough to be familiar with its flow and what one needs to say at each point.

2. *Room-related issues*: As stated earlier, a researcher should never see the location where they are to give a talk only shortly before they do so, and should be aware of any potential issues around screen location, distance to audience, vocal amplification system being used, ambient light, projector strength, and so forth, so that these issues can be identified early and mitigated where possible.

3. *Timing-related issues*: a key point is to make sure the time allocated for a talk is known and adhered to, which is basic preparation. However, I was caught on this once, not too long ago, when given a 15-minute (short!) slot for a research presentation, which I intensively practiced to make sure it fit within 13 minutes, assuming 2 would be left for questions. Around 10 minutes before my talk, however, I discovered from the chair that they wanted to leave five minutes for questions, and so without time to delete slides I needed to revise my "script" in real-time and drop a couple of explanations and points I had planned to squeeze in. This is not ideal!

4. *Technology issues*: as mentioned earlier, bringing your own remote control or even computer can give confidence in how and whether things will work, but still the occasional hiccup can occur. I gave a plenary talk in November 2017 in a large cinema-like auditorium in Spain in which there may have been 500 people present, and toward the end had five slides with specific data-laden graphs to illustrate some recent work from my group. I had checked through the slides on stage earlier that day, and all was well, yet when the time came these slides came out somehow missing the graphs, although other elements of the slide, including arrows and annotations, remained bizarrely visible. I looked to the side, where the audiovisual team for the conference were looking as flustered as I was, but I had no choice but to "improvise" somehow, apologize without blaming anyone, and blunder on. I came to my last slide and finished, I am sure less than spectacularly, then saw motion to the side of the stage where the audiovisual technicians were pointing to the screen, on which my first problematic slide now appeared, resplendent with the original graphs. I then looked to the chair of the session, who signaled that I could (re) proceed, and so I repeated the last few minutes of my talk—a very odd experience!

The Really Hard Bit: Answering Questions

The fifth thing that can go wrong, which is sufficiently serious to warrant a section of its own, is answering questions afterward. This is the one area where no amount of rehearsal can completely cover all possible eventualities, unless such rehearsals include an audience who have been primed to challenge the speaker with the widest range of possible approaches and questions that might be raised. The worst thing that can ensue when the chair of a talk declares it open for presentations is that the impression of confidence, competence, and expertise which the speaker has worked so hard to convey is thrown away by a flustered or bumbled response to a question.

Questions proffered by an audience can range from the intense to the insane, and not infrequently the main reason someone asks one is actually to make a statement or show off knowledge of their own, where the actual response required from the presenter is cursory. Other questions can of course be readily disposed of with clarifications or additional details from a well-prepared presenter who is on top of their topic. However, where a question genuinely raises a point not previously thought of, an alternative explanation for results, or some other perspective that somehow leaves the presenter at a disadvantage, the key is not to become embarrassed, argumentative, or defensive, but perhaps respond along the lines of that being something that is useful for future research, or very helpful for future considerations.

The presenter should not get into a one-to-one argument or debate with an audience member, and the chair should not allow this to arise, so the sense of a hot spotlight burning down on a hapless presenter stuck like a rabbit in headlights should only be brief, if painful. The key is to remain calm and dignified and, above all else, watch the seconds on the clock tick onward until the time allowed elapses.

Posters: The Less Scary Option

At perhaps the lowest level of the hierarchy of prestige in scientific communications lies (or, more precisely, sits stuck vertically to a wall or board) the poster, typically a young researcher's first foray into displaying their work and wares to a potentially interested audience.

A poster is a single sheet (usually) of A0 size, which is around 3 feet by 4 feet, usually of portrait but sometimes landscape orientation.[7] It is typically designed in PowerPoint, and in style represents a mix between a paper and a presentation, being more of a static exercise in reading than the latter but much more open to design and creativity than the former.

When preparing a poster, it must be borne in mind that two types of people will read any poster at a conference:

1. [The smaller group] Those who have seen the poster title on a list of posters and have a specific interest in the subject matter (or presenter) and make a bee-line to read it;
2. [The larger group] Those who are not specifically looking for that poster but are merely wandering around a room full of dozens or hundreds of posters, perhaps with a cup of coffee in hand, and are looking to see what seems interesting to read.

Members of the larger second group might have 30 minutes in which they hope to look at a bunch of posters and will typically wander up to posters that attract

---

[7] In my experience, landscape posters are quite common at conferences in the United States, but usually a portrait format is preferred as more posters can fit on a certain width of display space. The first thing to check when preparing a poster is which is expected, as this is a very embarrassing way to make a bad first impression.

them visually or by title and make a very quick decision (within one minute) as to whether to linger to read or wander on.

The poster must be designed with this type of viewer/reader in mind, and "sell" them a snapshot of the research in question very quickly, by maximizing use of simple (as possible, without undermining the science) visual images and plots and cut-down text. A useful learning exercise could be to simply observe browser behavior at a busy conference poster session, and watch which posters attract and retain the most attention from evidently casual passers-by. I bet it will be those that minimize dense blocks of text and have the most striking and attractive appearance, and then back this up with a credible and interesting scientific story to tell.

Normally, those who have prepared posters will have an allocated time at which they are expected to be by their poster, to answer questions or engage in discussion about the work shown. It is dispiriting to see those such presenters (remembering these are often students displaying their work for the first time) standing forlornly by their poster or getting briefly excited and nervous when someone comes up to look but then wanders off wordlessly after less than a minute. Posters are a great way to make contact during such conversations, though, and presenters should take the initiative to start discussions with anyone who stops by in such a case, noting that such discussions are very different in nature from those that ensue around a formal presentation, being more on a level playing field (both parties standing), with a limited audience and free time for contact (possibly positive or negative).

When designing a poster, whether landscape or portrait, a key principle is that a poster should have a clear flow so that the reader knows how to follow the flow of information, and a good guideline is that the flow should be vertically in columns of text or images, rather than horizontally. This is because, in theory, if a number of people were reading the poster at once, a vertical arrangement allows them to shuffle sideways as they read, without getting in each other's way.

There are a number of excellent resources available online for poster design, one of which was developed as part of a project in Ireland to create resources for postgraduate students.[8] This site includes a great range of text and video tutorials and tips, and half an hour spent there before designing a poster will lead to a lot of good ideas.

In Conclusion: The Limitations of Presentations as a Mode of Communication

As we have seen, presentations at conferences and seminars are a key element of a researcher's communication strategy and, while transient and limited in audience compared to a paper, serve many other functions. These include the generation of feedback, contacts, and opportunities as well as the benefits to researchers of

---

[8] http://www.nuigalway.ie/remedi/poster/index.html

taking a large block of work and chipping away at it until a coherent story is exposed.

While not wishing to end this chapter on too negative a note, however, there are some frightening examples of how poor use of communications tools like PowerPoint can have awful consequences. Specifically, while this may seem hard to believe, investigations of the loss of two space shuttles (Challenger in 1986 and Columbia in 2003) blamed, among other factors, endemic poor use of PowerPoint as a communication tool within NASA and partner engineering companies as a contributory factor to the loss of life that ensued. The point was that slide decks prepared during critical windows in which identified problems needed to be solved were not effective in communicating the problems or technical details involved, and so did not result in the correct actions and outcomes. There is a significant amount of information available online about this (an Emeritus Professor at Yale called Edward Tufte did a lot of such analysis), analyzing slides that showed hugely complex levels of information within four-level nested bulleted lists, unclear terminology and phrasing, and lack of attribution of key information to "owners" who could speak to and explain it.

The report of the Columbia Accident Investigation Board includes the following shocking paragraphs:

> As information gets passed up an organization hierarchy, from people who do analysis to mid-level managers to high-level leadership, key explanations and supporting information is filtered out. In this context, it is easy to understand how a senior manager might read a PowerPoint slide and not realize that it addresses a life-threatening situation.
>
> At many points during its investigation, the Board was surprised to receive similar presentation slides from NASA officials in place of technical reports. The Board views the endemic use of PowerPoint briefing slides instead of technical papers as an illustration of the problematic methods of technical communication at NASA.

The overall lesson from this tragedy is that, as a proper medium of communication of technical information, PowerPoint slides will never be as effective as properly written reports.

Now, on that sombre note, we will change gears and look in the next chapter at the techniques and approaches that might be used in communicating with nonspecialist audiences, including the general public, the media, and other stakeholders in research.

9

Expanding the Comfort Zone

COMMUNICATING WITH NONSPECIALIST AUDIENCES

As should be clear from reading this book so far, a key requirement for all research-ers is to be able to communicate their research in a formal and highly precise way such that their peers can evaluate their arguments, methods, and conclusions, and then subject them to detailed criticism and interrogation. Whether submitting a paper to a peer-reviewed journal or presenting at a major international conference, the expectations are of objective, calm, unemotional, and rigorous presentation of the facts.

Scientists are trained in these skills and judged for them from the start of their career. They may undertake training in academic communication, learn from hard experience and the rigors of peer review, and perhaps occasionally read books such as this one for advice on how to do it better. This is all driven by the hard fact that progression in a research or academic career will likely depend on one's success in these activities.

However, today, even mastering the skills of professional communication is not enough. It is increasingly recognized that successful researchers need a lot more tools in their toolbox than those required for high-level communication with expert peers.

Every form of communication must start with the principle of considering the audience, but the modern researcher must also be prepared for the widest possible range of audiences, and be able to tailor their message and means of preparation accordingly.

For example, a researcher working on the impact of a certain treatment on hepatitis might have a busy week where they speak on a Monday to an expert audi-ence of researchers working in that very specific field, who are deeply familiar with the tests, terminology, and mechanisms involved. On Tuesday they might speak to a group of medical professionals whose primary focus is on the implications for their practices, on Wednesday to a group of patients worried about the impact of the disease on their lives, and on Thursday to a journalist for a major newspaper writing an article for the general public on progress in the fight against the disease.

Finally, on Friday, they may then be focused on writing up their findings for publication in a specialist peer-reviewed journal.

Would they use the same language and approach for each audience?

In another example, a researcher studying the impact of a certain fertilization treatment on countryside land use could be regularly speaking to experts in the (ahem) field, farmers wondering about the implications for their practices, and policymakers from government focused on policy and regulatory implications.

Every researcher is likely to be able to articulate several such scenarios for their own topic, even before we get to the classical situations like explaining it at home to your family or to the archetypal "man in the pub."[1]

There are some very nice videos online from *Wired* magazine in which experts are challenged to explain complex concepts at five different levels of difficulty, as in to a child, a teenager, an undergraduate student in the area, a postgraduate student in the area, and finally a peer or colleague. These show clear examples of how a single message can be tailored in many different levels for different audience.

Researchers need to be able to work out which communication strategy is likely to be most effective in each circumstance.

What are the advantages of being able to communicate effectively at multiple different levels?

1. It has been said[2] that you don't really understand your own research until you can explain it to your grandmother. There is no doubt that having to boil your research down to key understandable messages and points is extremely difficult if you do not completely understand it yourself. So, tackling the challenge of communicating to an unfamiliar audience forces a level of examination and reflection which can have enormous benefits for a researcher's engagement with their own topic. In addition, the questions asked are likely to be very different to those from a specialist audience, and further force researchers to consider their work from unfamiliar angles and think of aspects that might not have previously been apparent, and that can be hugely illuminating and positive.

2. As discussed throughout this book, a key objective of any researcher should be for their research to have impact. For most areas of research, impact can happen in many ways and on many levels, and examples such as influencing public opinion on a controversial topic, changing doctors' practices to improve patient care, or influencing government policy can be as important as publishing a paper in a prestigious journal that remains inaccessible or unread by nonspecialists.

---

[1] There is in Ireland a program of talks called "Pint of science," which seeks to test this possibility, scientifically of course.

[2] Commonly, but controversially, attributed to Albert Einstein, but then again many quotes are.

So, researchers need to be able to design communications strategies for a range of scenarios. How do they do this?

As with every communications task, the starting point in communicating with nonspecialist audiences is the audience themselves:

- Who are they and what are they interested in?
- What do they know to begin with (point A)?
- What do you want them to know at the end (point B)?
- Based on who they are, how can you help them to get from point A to point B, via a journey that is rewarding and interesting for them, as effectively as possible?

The question then is, what modes of communication are to be used? This could be a written article (for a website, magazine, or newspaper) or face-to-face communication, such as a presentation or the multitude of other forms and fora of direct communication that exist today, such as festivals, theaters, social media, citizen juries, and more.

As with all types of scientific communication, the difference between preparing communication where the communicator is not present when the audience receives the message (written articles) and where they are (presentations or verbal communications) is critical, as different skills are needed, and also different levels of completeness of the information presented. The different considerations for both modes of communication will be considered next.

In many universities worldwide, it has been increasingly recognized that the ability to communicate effectively to nonspecialist audiences is a key skill for doctoral students to acquire and graduates of advanced degrees to be proficient with. It is also acknowledged today as a key activity for academics in general to engage in, in terms of enhancing the reach and impact of the work conducted therein, much of which will have been funded by the taxpayers. Indeed, funding agencies will increasingly ask for nonspecialist or lay communication strategies to be outlined as part of proposals seeking funding for research, and then use the number and nature of such activities as metrics on the basis of which to later evaluate a project's success.

For all these reasons, in my own university we introduced a number of initiatives in the last decade specifically to help students to develop these skills. A key pillar of this activity was a journal (*The Boolean*, theboolean.ucc.ie) in which PhD students shared the outcomes of their work in lay terms, and where acceptance of submitted articles was purely based on the ability of someone completely nonexpert in the topic in question to understand the work and its significance. The second main activity was a Doctoral Showcase, where the students competed in front of a panel of judges from the media and wider society to demonstrate their flair for explaining their research through 10-minute presentations, a "3-minute thesis" format, or a creative mind category, where any possible nonacademic communication medium could be used. Some of the entries in the Boolean and Showcase demonstrated astonishing proficiency in creative nonspecialist communication, and I will refer to some examples of these in what follows.

A former student of my university with a superb ability to communicate with nonspecialist audiences once said that, if the audience in such a scenario does not understand what you are saying, it is not their fault, it is yours, and you need to try harder. This is a key message that cannot be forgotten, whatever the type of communication concerned.

Writing for the Nonspecialist

For writing outside the traditional realms of the peer-reviewed paper or expert audience, a whole new range of techniques becomes available for use. This is in part because these methods are more likely, if used well, to successfully engage a less expert audience, but also because the writer can shake off the sometimes fusty confines of traditional academic writing, think laterally, and embrace their audience's viewpoint.

This includes things like humor, analogy, storytelling, rhetoric, quotations from nonscientific sources, questions, and many more. On the other hand, it excludes a wide range of terminology and words that might be key for use in academic discourse.

There is no doubt that formal scientific communication uses a range of words and constructions that a researcher would not use outside research-related discussions and which they leave aside once they rejoin "normal" conversation in their day-to-day lives. These words must similarly be left aside when communicating with a nonspecialist audience. Obviously, these will include specific technical terms and jargon specific to the topic in question, but also the dry and formal phrasing that is used to convey the objective tone expected of a scientific paper but that will sound stilted and unengaging for a different audience.[3]

A key difference between writing for the specialist and nonspecialist that can never be forgotten is that the latter audience is less "captive" than a traditional scientific reader, who often must read something solely because it is relevant to their own work. In contrast, the nonspecialist reader is reading for pleasure or interest, without compunction to finish the piece being read.

Every researcher can identify with struggling through a dense complex or plain boring piece of text because it is necessary to do so as it is critical for their work in some way, and so the end will justify the means.

---

[3] In my own writing and speaking about food for a general audience this has been repeatedly driven home in terms of how, if we use a scientific, technical, or functionally descriptive term for a food ingredient, it elicits a completely different reaction than if we use a more familiar term for exactly the same component. "Egg," "vitamin," "starch," and "honey" are viewed by many very differently than the words "emulsifier," "antioxidant," "stabilizer," and "humectant," for example.

In contrast, any lay reader encountering a piece in which the writer has failed to justify their attention will simply lay it aside and move on, as we all do with article in magazines, newspapers, or online. The responsibility lies with the writer to grab, maintain, and deserve the reader's interest and attention. If the reader does not understand, the writer must try harder.[4]

I certainly found this when I turned to my book *Molecules, Microbes, and Meals: The Surprising Science of Food*, where my goal, when trying to write about a wide range of scientific topics in the chemistry, microbiology, and processing of food, was "accessibility by any means necessary." In that case, I used as much analogy, humor (some highbrow, some low-), cultural references (ditto), light political humor, and above all else striking imagery, both verbal and literal, with beautiful original photographs of the outer (familiar) and inner (microscopic) appearance of food being a key way to achieve the objective of presenting food from unfamiliar angles and hopefully causing the reader to look at it in a new light.

For any written piece or article, everything the reader needs to understand the topic in question needs to be on the page, as the author is not present to explain or answer questions, but on the other hand the reader can pause at any time to check a fact or definition or look something up, if they are confused or unsure. However, this presupposes that the reader has been sufficiently "hooked" not to simply abandon the task at that point, and ideally understanding the piece should not depend on some additional research and cross-referencing on the readers' part.

Grabbing their attention starts at the very beginning—the title. The title is an immediate opportunity to captivate and tantalize the reader. The best titles for such pieces are unlikely to be the same as those which would appear atop a corresponding peer-reviewed article. Jokes, quotations, questions, or provocative statements can all be key weapons in grabbing the reader's attention, like a window display or sign that draws a meandering shopper through the door into a particular shop.

Let's consider a scientific paper with the dry title "Evidence for Residual Extraplanetary Biological Activity in Archaic Basalt Deposits." How would this be translated into a "popular" article title? Perhaps "Alien Life in Ancient Rocks." A researcher writing about his work on supercold physics at temperatures close to those (absolute zero) at which the movement of molecules slows and even stops would surely be tempted to refer to "how cool is that?" or some such cultural allusion or pun to break the (presumably supercold) ice with the reader, or challenge them with a question about the coldest place in the universe. This approach can also be seen when a newspaper article on a strange form of life called a sea cucumber, which has the scientific name *Enypniastes eximia*, referred in its headline to its more interesting name: "headless chicken monster."[5]

Overall, the title must engender a sense of interest, curiosity, and even excitement. All that training in being calm, objective, and professional in scientific

---

[4] To avoid the pitfall highlighted by George Bernard Shaw when he said, "the single biggest problem in communication is the illusion that it has taken place."

[5] *The Guardian*, October 21, 2018.

writing must be suppressed, possibly forcefully, to allow creativity, humor (groan-inducing puns being a frequent feature) and even a little hyperbole, without making promises the article and research cannot then keep, to drive the approach. Researchers need to be interested in and enthusiastic about their own work sufficiently to keep them motivated through the long hours and frustrations that are inevitable in every area of research, and popular writing allows them to express this openly for once.[6]

Having got past the title, the first few lines of the piece itself will be critical, in terms of building on the promise of the title to draw them further into the topic. Going back to the idea introduced in chapter 8 of zooming, the start could perhaps link directly to the reader's known experience and situations or ideas with which they can readily identify, to ground them in familiar territory before taking them by the hand and leading them gently into the unknown. This could be achieved by using a question, analogy, or popular reference, or an easily pictured hypothetical scenario.

One article in the Boolean journal mentioned previously,[7] on the topic of refugee camps in sub-Saharan Africa, started with the fantastic line "Let's close our eyes for a minute and imagine that the Ireland we know today doesn't exist." The author then goes on to speculate about an alternative Irish historical timeline, where developments mapped those of that region of Africa, featuring war, famine, coups, failed peace treaties, and religious extremism. Having grabbed the (Irish) reader's attention and put them briefly into the situation he is about to describe, he then proceeds through a series of sections with attention-grabbing headings like "Dadaab . . . where the hell is that?' and "Sorry, we're full!"

In any article aimed at the nonspecialist, we must lead the reader, word by word, into the thickets of the work in question. In this uncertain land lie traps and pitfalls that can deter or despair the reader, and most of these take the form of words that are familiar to the specialist or expert, but seem obtuse or intimidating to others. These are like secret codes or shibboleths[8] that mark the border between those who know the field and those who don't, like words in an unfamiliar language, which, if not translated, will make a text hopelessly impenetrable. The same applies, perhaps even more so, for the use of acronyms and abbreviations, which are commonplace for the expert but can appear remote and intimidating for others.

In all cases, the researcher writing for a nonspecialist audience needs to identify these terms and either defuse them, by explaining them clearly at the first instance

---

[6] While scientific writing is an art and skill of its own, as discussed throughout this book, it is hard occasionally not to think of Truman Capote's withering putdown of some other writers who lacked what he regarded as literary style: "That's not writing, that's typing."

[7] Damien McSweeney: "Dadaab . . . A Forgotten City in the 21st Century."

[8] A custom or principle that distinguishes one group of people from another, and a lovely word I first discovered through the magnificent TV show *The West Wing*. While frequently associated with religions, science with its various tribes is full of shibboleths that form barriers to those who are nonscientists.

of use, or remove them entirely, if this can be done without making it impossible or highly awkward to discuss the topic at hand.

The manner in which a piece of text is laid out can even influence the manner in which a non-specialist reader will regard it. Long paragraphs of dense text look intimidating and boring, and a few deftly applied paragraph breaks can be a simple way to make text look more approachable. Dividing up the text (which should likely not be too long anyway) with regular section headings (which should take the same approach as the article title, aiming to be provocative or capture the reader's interest) will also be helpful in conveying the idea of "breaks" in the text, like rest points for the reader's attention and effort.

Images are also a great way to make the text look interesting or enticing, probably not complex diagrams or figures but rather striking images or photographs, or perhaps graphs. These would not be like those which might accompany a more formal presentation of the work but have been reconfigured and redrawn to show a simple trend or effect.[9] If properly done, this gives a reader a sense of "seeing the science" and not being patronized, as they are being trusted to interpret the basic data shown, but this is done at a level that is carefully calibrated to meet the level of the expected readership. The same can apply for reasonably simple tables, which could summarize information in a much more effective way than trying to do so in the text.

To find a masterclass of writing about science for a nonspecialist audience, there is no better example than Bill Bryson's book *A Short History of Nearly Everything* (2003). Bill Bryson was always one of my favorite writers for his travel books, blending beautiful writing with great humor, so, when I heard he had written a book in which he sought to tackle the entire scope of scientific knowledge, I was deeply excited, and the book does not disappoint. Every single page of this magnificent book is a master-class in how to make complex facts accessible, presumably reflecting the work Bryson had to do, as a nonscientist, to master such a range of topics himself. As mentioned earlier, you really need to understand something to be able to explain it well to others.

To take one example, in his first chapter ("How to build a universe"), he starts wonderfully as follows:

> No matter how hard you try you will never be able to grasp just how tiny, how spatially unassuming, is a proton. It is just way too small.

He then goes through a series of comparisons of the size of a proton, directly speaking to the reader ("Now imagine if you can (and of course you can't). . . . Excellent. You are ready to start a universe") and drawing them in as if having a conversation with them alone. In the second chapter of the book (which he bridges into with a lovely linking phrase of "before we turn to the question of how we got here, it might

---

[9] Great examples of this approach can be found in the documentary *An Inconvenient Truth* and its sequel.

be worth taking a few minutes to consider just where exactly 'here' is"), he master-fully gives a sense of the vast scale of the solar system. He achieves this using a wide array of analogies and familiar benchmarks (Earth as the size of a pea, on which scale Pluto would be 2.5 kilometers away from it, and the nearest other star 16,000 kilometers distant, and the astonishing fact that Pluto is one-fifty-thousandth of the way to the actual edge of the solar system). Every successful tactic in bring-ing complexity down to a relatable level is used, from analogy, gasp-out-loud facts clearly expressed, a sense of direct communication with a caring author who is com-mitted to making you understand, to a frequent wry smile or even outright laugh, and everyone who wishes to write well on scientific topics should be made to read this book.

When I was younger, books like this were a huge part of making me want to be a scientist,[10] and they continue to remind me why I want to do what I do. When I reflect on what some of my favorite books on science conveyed to me and how they cast their spell, key features included a dramatic sense of excitement and of the personalities behind the discoveries.

For an example of the former, Horace Freeland Judson's book *The Eighth Day of Creation* (1979), my paperback copy of which I have just taken off my bookshelf and found to be almost falling apart from multiple rereadings over the years, describes the early era of discoveries in molecular biology like a thriller. The steps leading to Watson and Crick's completion of their double-helix model is presented in a series of sections, some very short and almost breath-lessly punchy to keep the story moving ("The drums were beating. On January 6, [Rosalind] Franklin wrote to [Robert] Corey to ask for details about the struc-ture. Corey did not reply until April 13"), with great lines like how Sir Lawrence Bragg, their boss, at one point told them to stop working on DNA, "with iron showing beneath the tweed."

Indeed, the more famous book on this discovery, by James Watson himself (*The Double Helix*) was an immediate and lasting success specifically because it related the tale of the discovery in in a novelistic manner, gossipy (in a famous description of Rosalind Franklin, he writes, "And yet I could not regard her as totally unin-teresting. Momentarily I wondered how she would look if she took off her glasses and did something novel with her hair") and controversial (to put it mildly, par-ticularly in terms of the details he revealed for the first time about the influence of Franklin's data on the model he developed with Crick). It plunges the reader into the heart of a few critical months in Cambridge, and does the scientific equivalent of showing exactly how the sausage is made, capturing excitement, competitive-ness, rivalry, and the synergistic boon of effective collaboration, particularly when

---

[10] Although I was the youngest in a large family of nonscientists, It seems I was infected by the science bug at an early age, perhaps by a random mutation in an unidentified gene, and I was mak-ing laboratories out of Lego at an age where cars and castles might have been more common among my peers.

contrasted with the problems that result when researchers cannot work together, for reasons such as clashing personalities.

Another much-read book on my shelf, *Lonely Hearts of the Cosmos* by Denis Overbye (1991), describes several decades of research in the fields of cosmology and particle physics in a beautiful manner[11] by, in particular, including pen portraits of each researcher introduced, from their clothes and offices to their manners and speech patterns, completely humanizing their work (as Bryson also does).

In many ways, this is the key objective of good science writing, which is to change the perspective of science from a product to a process. When we are presented with science through "hard" channels, such as school or college or textbooks, the focus is on the outputs and what is known, and far less frequently on the process by which these things came to be known. The failures, frustrations, and human experiences that lie behind every single published line of objective statement and argument are rarely glimpsed. Science is the way in which we come to know about the world, and is dependent on the activities of people. To bring the nonspecialist into the world of science, we must turn the focus from the output to the process, and provide avatars to guide them. This could be through the writing of the author engaging them companionably, as Bryson does so well, or through making them feel like they can picture the scientists concerned in their mind's eye, observing them as they go about their work and helping the reader to understand what they have done and found.

This is also the case in David Quammen's more recent *The Tangled Tree*, a history of the development of the understanding of the kingdoms of life and how molecular biology in the 1970s overturned conventional wisdom regarding the main families of living things by identifying, besides eucaryotes (plants and animals) and procaryotes (bacteria) a third kingdom, the archaea. Quammen's beautiful book makes the science come to life through the stories of the protagonists, stretching back as far as Darwin, and displays the rousing power of a great analogy in explaining a complex construct. I must admit to putting the book down on reading his description of the action of ribosomes, the cellular factories that translate the DNA code into proteins, as being, due to the complex shapes that the proteins thus created adopt, like that of a 3D printer, and exclaiming aloud with jealous admiration at the cleverness of the comparison.

Recently, public interest in the announcement of the first photographic representation of a black hole in April 2019 was almost certainly enhanced by the identification of this story with the image of a female scientist at Harvard University who was deeply engaged in the development of the image-processing software used; images of Katie Bouman's contagious enthusiasm and excitement invariably accompanied articles on the discovery and gave a human face

---

[11] My copy bears a quote from a review in *New Scientist* that "it's a rare science book that can make you cry, but, I swear, this one brought me close," and it brings you very close indeed.

and emotional resonance to a story over which many readers might not other-wise have lingered.[12]

In the case of any piece of writing aimed at a general audience, whether long or short, at the end of the piece, a clear final thought or perspective will be key to ending on a "bang." Rather than leaving the reader with a "so what" sensa-tion, the writer should aim to leave the reader with a clear sense of having learned or understood something new and with a resulting feeling of satisfaction. They should zoom gently back out to the real world, to which the reader will implicitly return on completing the piece, by leaving with a final link to experience or future behavior might be very effective: "So, next time you see/hear/notice . . . then you will . . . ."

Presenting for the Nonspecialist Audience

Another key means of explaining research will be through oral presentation, whether to a public audience or group of interested stakeholders.

We have discussed the key steps in preparing a good and effective presenta-tion in chapter 8, and many of the same principles apply. Nonspecialist presen-tations will be constructed based on a mixed delivery strategy, combining vocal explanation and audiovisual backup, or sometimes will be verbal only. In either case, the verbal elements are arguably even more critical, presenting the chance to argue, provoke, entertain, amuse, and above all persuade. The key objective of the speaker is to make the audience interested in the work to begin with and keep them sufficiently interested to listen to what the speaker has to say.

Compared to a traditional scientific presentation, the audience has no sense of duty to struggle through a talk because they somehow "need" to hear what is being said, and where necessary suffer through bad delivery because the message is more important than the medium. The nonspecialist audience is, once again, not captive, and are purely there for interest, thereby being free to leave, whether physically or mentally, at any time.

The speaker must convey a sense of credibility, and the audience must trust them, but there is even more room to try and make the audience "like" them, as empathy with the speaker in such contexts will absolutely enhance engagement with the material.

---

[12] Searching for jokes that appeared on Twitter around the discovery will yield some great examples of how to find humor in any scientific discovery, which is so important to making any discovery accessi-ble to a broad audience. I particularly enjoyed the analogies to Brexit ("an all-consuming blackness that is sucking in everything around it and destroying it"), and those where the black hole was unimpressed by the quality of its image, and asked to be untagged from the image. The identification of Bouman with the discovery did, however, also showcase the darker side of public interest in science, as it led to some very unpleasantly sexist commentary.

The analogy in chapter 8 of the remote control is even more critical for the non-specialist audience, as they cannot pause the speaker to check something, rewind to catch up on something they have forgotten or missed from an earlier point, or fast-forward someone who is boring or telling them things they know already. The only button they have is the physical or mental eject option mentioned earlier.

The points made for written articles about selection of language and use of arresting headings and questions applies equally, if not even more so, to the oral presentation. Grabbing the audience's attention is key, and this also extends to the use of gesture and eye contact, to engage directly with the listeners. If anything, theatricality and drama, as well as projection and oration, are even more critical for such cases than for a conventional academic format.

Turning to the audiovisual side of the presentation, if this is used, slide design will be driven by different principles for the nonspecialist. Images (the more striking and bold the better) will have far more impact than text (which should be minimized) and, as for written nonspecialist communication, graphs and tables will (where needed and used—and they should not be shunned completely) be stripped down to key extractable messages. Without compromising on the integrity or credibility of the speaker, which need to come through to make sure the message is taken seriously, there is probably more room for humor, cartoons, or unusual image juxtapositions than might be considered for a scientific talk. Animation, or video clips, can also avert possible ennui by livening up the presentation.

Just make sure that the main take-home message remembered days or a week after a talk is not the joke or cartoon!

Other Strategies for Nonspecialist Communication

We have focused thus far on written and oral presentation of scientific communications, but beyond the traditional "talk" structure, a wide range of communication formats have been tested and used to challenge and engage broader audiences.

For example, to encourage researchers to distill their work and findings down to something closer to their core message (or small number of messages), a common format is a "three-minute thesis," where the speaker has, yes, 180 seconds to present their information, perhaps supported by a maximum of one slide. This format is also sometimes referred to as "lightning talks." One way to consider this is to keep asking "so what?," until the message of the research has been distilled into a raw take-home message.

A particularly tortuous extension of this is called Pecha Kucha, which apparently originated in Japan, that gives the speaker just under seven minutes to present their work, but on the basis of using 20 slides, which are timed to change and advance (without the speaker intervening) every 20 seconds (giving a total time of 20×20, or 400 seconds, a brisk total of 6 minutes and 40 seconds); the speaker is required to shut up and sit down once the last slide disappears. Compared to the "one slide equals one minute" principle common for normal presentations, this is a breakneck speed, and while not feasible for presentation of detail on each slide, can lead to a fast-paced energetic run through more top-level information. I have seen this done very well, and thoroughly enjoyed it being done rather less well, where a speaker is still talking about something that appeared three slides ago. Around a decade ago, an even crueller variant emerged called Ignite, in which the time per slide is cruelly reduced to 15 seconds (total talk time of 5 minutes).[13]

Going even further, high-profile guest speakers at the IgNobel Prize awards, given annually for research that first makes people laugh, and then makes them think (see the website Improbable Research at www.improbable.com), have to give 24/7 lectures about their research, in which they explain it twice, once as a complete technical description in 24 seconds, and once as a clear summary that anyone can understand, in seven words.[14]

Taking a very different approach, in the United States in 2007 a competition, sponsored by *Science* magazine and the American Association for the Advancement of Science, was launched to encourage researchers to "dance their PhD," by designing a piece of interpretive dance based around their research. Finalists in the 2017 competition, which had 53 entries, presented through this medium research on topology, biochemistry of forensic science, the ecology of sea stars, and the psychology of creativity. There are plenty of videos of such performances on YouTube,[15] and they are rather dazzling, but I do wonder whether, if watched without any knowledge of the student's actual research area, an audience would be able to predict this from what they have seen.

In my own university, we had a "creative mind" category within our Doctoral Showcase competition that challenged students to effectively present their work to a lay audience by any means possible. Over the years, this resulted in work being presented through mime, puppetry, and joint presentations between students on different continents. One of the most dramatic moments of all, however, was when a student described his work on legislation around prostitution in Ireland by first showing a video he had prepared on the background to the work, including clips of political debates; dramatic music swelled toward the end, and then he appeared and walked the length of a large formal hall while singing about his work and methods in the style of a West End or Broadway musical, with the talent to match, finishing up singing at full throat to the audience from the front of the hall. That was a presentation which few who saw it will ever forget. Everyone learned

---

[13] The name may come from an occasional frustrated presenter spontaneously combusting with effort.

[14] One example, from a real Nobel Prize–winning professor of economics at Harvard University, Eric Maskin, on uncertainty: "Uncertainty is the only sure thing—perhaps." Another comes from a researcher studying reproductive behavior of fireflies: "Female fireflies favour fancy food-filled flashers."

[15] This reminds me of a recent illustration of the way in which language and science are rapidly changing between generations, where, in preparing for a course, an experienced visiting academic asked for a U-tube viscometer, which caused some confusion with younger students assisting, who thought this might be a new revolutionary type of online or inline measurement system.

something new about this topic in the process, and that learning was embedded indelibly by the force and skill with which it was performed.

However, as with the effective use of humor, such bold strategies can only successfully be delivered where the speaker has the confidence and ability to commit fully and do it justice, as the risk of embarrassment if something falls flat is proportional to the level of creativity and boldness involved.

Communicating with the Media

In terms of communicating the outcome of research, a small proportion of topics will be deemed (sometimes for unpredictable reasons) of interest for dissemination through conventional (as opposed to social) media, whether print (and inevitably online) newspapers, radio, podcasts, blogs, or television shows. This will typically apply to research believed to be of public interest or timely, while work will occasionally be selected because it was somehow found and seen to be quirky or humorous.

As opposed to researchers writing or presenting their work themselves, in such cases it will be filtered in some way through a journalist or other media professional, who takes responsibility for how it will be presented to the world.

Researchers' interactions with the media may be of a number of different types:

1. Pro-active research promotion: all large research-active universities and
    institutions today have media offices who eagerly seek opportunities
    to bring work from that organization to the public eye in a way that
    maximises visibility, and will work with researchers to craft press releases,
    news conferences, and schedules of interviews. All recent major discoveries
    (such as possible life on Mars, the Higgs Boson, and gravitational waves)
    have seen the formal announcement through peer-reviewed publication
    mirrored by a coordinated media strategy, and indeed most journals will
    have clear procedures and protocols governing when such announcements
    may be made relative to the timing of publication, so that neither formal
    scientific nor broader media dissemination activities are seen to "scoop" or
    interfere with each other.

    Crafting a good press release can be a very different form of writing
    than any other form of scientific communication. Researchers should
    draw on the experience of professionals here, to identify the small number
    of key points that need to be put in writing as clearly and unequivocally
    as possible, blended with the right level of context and drama to make
    journalists interested, and draw them into the space where they see how to
    "sell" to the public the messages that form the core of the piece. It has been
    frequently noted how many media outlets will publish press releases almost
    verbatim, and so the responsibility and opportunity can lie firmly here with
    the researchers to present their case directly to the public.

2. Reactive research explanation: in other cases, researchers may publish papers (or reports) without the fanfare of major cross-platform multimedia extravaganza and find that some journalists have come across their work and wish to write about it, and hence seek comment. In such cases, the researchers must be very clear what messages they wish to get across, and what the benefits of such interactions might be. In my own work, for example, I have consistently declined interview requests on one area of research (the composition of human milk), for the reason that the topic could be quite controversial or emotive (a key draw for media attention), and my team's interest is in the science, but yet to make a good story will require the identification of an "angle" or controversial comment. This would not benefit perception of our work and could easily offend or anger any of several different cohorts of stakeholders interested in that area. On the other hand, I have happily been interviewed for hard-hitting topics like 3D-printed cheese and the science of Christmas food!

3. Commentary on topical issues: researchers could be deemed to have a public responsibility to comment on or advise the public when something of broad general interest, in an area in which they have expertise, arises. For example, a researcher might get a call when a scandal or controversy in a different country related to their field of research hits the headlines, to help the journalists put it in context and explain the background, either through "behind the scenes" advice or through on-the-record interviews or commentary (I have done both). There are often sites and lists maintained by institutions where researchers can make themselves "available" for comment on topics close to their own fields.

4. Ongoing information on broad scientific topics: some researchers have built very good platforms for public dissemination of science through newspaper columns, online blogs, podcasts, or TV shows or slots. These include physicists and cosmologists such as Brian Cox and nonscientists including the comedian Dara Ó'Briain and the actor Alan Alda, who established the Alan Alda Center for Communicating Science, which sets challenges such as competitions for scientists to answer, through written or video submissions, technical questions at a level specially aimed at an audience of 11-year-olds.[16]

All of these forms of communication, if they work well, should work to the benefit of the researcher and the public, and help to bridge the gap between science and wider society. The salaries and research of researchers in public institutions are ultimately funded in large part by taxpayers, and so it is only right and proper to engage in communication with them. This could apply to any of the forms of

---

[16] This started when Alda remembered pondering the question "What is a flame?" when he was 11 himself, and subsequent questions, put forward by school children, asked the same of time, color, and sleep.

nonspecialist communication referred to in this book but, as one of the key ways in which much of the public gets its daily information about all aspects of the world is through the media, it can be argued that interaction with the media must be seen as part of a researcher's responsibility and portfolio of communication channels. For many researchers, however, it might be rare (or never) that a journalist asks them for comment or they have a chance to turn their work into "breaking news."

In dealing with the media, key considerations for researchers concern the control (or lack thereof) of the final message, and being absolutely clear on what they are prepared to say or be unequivocal about.[17] The key, as for all forms of communication, is to consider the ultimate audience but, in many cases, there is a filter in the middle, where the journalist will parse or rephrase the researcher's words, or in interview ask questions that might drive the researchers into areas where they are not comfortable or prepared to be as dramatic as the journalist might wish.

The key consideration in managing the transactional relationship between researcher and journalist is to understand the motivations and wishes of the media. This concerns the story they ultimately wish to tell and the angle from which they intend to cover it, and the researcher must understand and accept the risks and benefits of the transaction, as well as the message they wish to get across and the manner in which they wish to be perceived by the audience. Researchers should aim to come across as confident and knowledgeable, but prepared to defend their work as firmly and politely as possible, and never so eager to please that they make simplistic or misleading comments.

Major discoveries of recent years have become increasingly accompanied by carefully choreographed media strategies. After the paper that controversially claimed to offer evidence of life on Mars was submitted to *Science* in 1996 and accepted for publication, although the journal and authors alike were anxious to maintain secrecy through the review process,[18] it was rumored that the text of the paper had leaked and was available online despite an embargo from the journal, and a press conference (followed the same day by an even bigger announcement, from the president of the United States, Bill Clinton) was organized. A statement from the NASA administrator claimed that "NASA has made a startling discovery that points to the possibility that a primitive form of microscopic life may have existed on Mars more than three billion years ago," a sample of the hyperbole with which the news was presented. At such fora, the scientists involved got their first taste of what turned into a major onslaught of skepticism of their work, which eventually led (as mentioned in chapter 6) to it today being largely disregarded.

Several discoveries that have later proven to be false or questionable have been announced at press conferences, which maybe just reflects that these are high-profile discoveries that attract a lot of attention, across media, the public and

---

[17] Often someone other than the journalist can select the headline to be used, though, and this can then completely change the tone of a piece.

[18] As described in Goldsmith, Donald. *The Hunt for Life on Mars.* (New York: Penguin Books, 1997).

scientific peers, and so any failures equally attract disproportionate attention. In a famous example of this, two chemists from the University of Utah, Stanley Pons and Martin Fleischmann, announced at a press conference in 1989 that they had evidence of nuclear fusion, at room temperature (so-called cold fusion), whereas that reaction was more normally found in the considerably warmer environments of stars and nuclear explosions. This would have opened incredible possibilities for cheap energy generation, and was greeted with huge excitement, but the work turned out to be error-filled and completely irreproducible.[19] While it is generally believed the researchers did not intend to deceive, the case is held up as an example of not rushing to the media with wild claims unless they can withstand the harshest scrutiny.[20]

In other cases, such as the fears raised about the safety of the MMR (measles, mumps, and rubella) vaccine, the impact of the now completely discredited study led by Andrew Wakefield was greatly magnified, with well-documented public health consequences, by a press conference announcing the results.

In more recent years, we have seen very carefully organized media launches of discoveries such as the Higgs Boson and gravitational waves, even in advance of publication. In the case of the former, the discovery was announced at a special seminar on July 4, 2012, in the huge laboratory where it was discovered, CERN, which was shared by live-streaming with audiences around the world online. This was then followed by publication of the findings, following refinement of the preliminary results presented, over that summer.

In the case of gravitational waves, a press conference was organized in Washington, with simultaneous meetings and press conferences, as well as many watching by live-streaming, around the world, after the paper describing the results had been accepted by *Physics Review Letters*. The press conference was scheduled to take place on the same day as the publication of that paper (February 11, 2016), although rumors as to the nature of the announcement had started to circulate in advance. A Caltech physicist who directed the lab (LIGO, Laser Interferometer Gravitational-Wave Observatory) at the center of the discovery created headlines and soundbites in the media with his impassioned and proud cry of "Ladies and gentlemen, we have detected gravitational waves. We did it!"

I myself have relatively limited direct experience of dealing with the media, my area of research not being quite at the Higgs Boson level of public interest, but in general I have found articles to which I have contributed to be fair representations of my work, although in almost every case there will be at least one line where you think, "that's not what I said," or, "that's not what I meant." In the

---

[19] There is actually a *Journal of Irreproducible Results*, which is a science humor magazine, well worth a look for great examples of science humor (in which rarified category I also highly recommend the PhD Comics website).

[20] The editor of *Science* actually said to the authors of the 1996 paper claiming evidence of life on Mars that they didn't want a repeat of cold fusion, which is why publication and announcement were closely coordinated and embargoes on early revelation put in place.

previous chapter, I mentioned experiences of media and conference interest in a paper on 3D-printed cheese, and several articles appeared on media sites internationally relating to this, for several of which quotes from me were included that must have been extracted psychically, as I had never been asked for any! Journalists will rarely give researchers right of approval of the final article (generally due to short deadlines) and so this is a risk that must be taken but, in every case, I have found that the overall benefit of the exposure outweighs any minor accompanying annoyances.

I have also recently had quite a number of interviews, mainly for Irish radio, in relation to my book *Molecules, Microbes and Meals*. My university circulated a press release highlighting several aspects of the book, but the one that drew the most media attention was the one chapter out of 18 relating to food processing, and attention focused on my being invited to defend why processed food might not be as bad for you as you might think. This led to some interesting and lively discussions!

Based on my experience, some useful tips on dealing with the media might be summarized as follows:

1. Give it time and attention: to present your work or answer tricky questions, often over a phone, in a few minutes, knowing that you might not see the result before the rest of the world, takes focus and concentration. Clearing your mind and some time (and finding a quiet place) to do this is a safer plan than trying to do so while rushed or distracted.

2. Be prepared to be rushed, as deadlines always seem to be imminent, and failing to respond to a request may easily result in an opportunity being missed when the journalist moves on to someone else who happens to be available.

3. Prepare well, in terms of having a few notes ready with key messages you want to get across, or perhaps specific quotes you wish to offer, as these are the kind of thing that otherwise tend to occur to an interviewee only after the interview has been completed and the piece has been filed for publication.

4. Aim to build a good relationship with the journalist, whether short-term or long-term, as they are more likely to represent well the words of someone who engaged with them in a friendly and collegial manner than someone who has a hostile or patronizing manner. Here the journalist is the audience, and the researcher must work to get them on their side just as if they were any other audience, while recognizing that the relationship must be professional and that the journalist has their own motivations for making the dialogue work.

5. Consider the broader audience: during an interview, the primary audience is the journalist, but invisible and silent behind them sits the constituency to which they will report. Whether this is the general public (or a specific demographic portion thereof, in terms of the readership of the particular

media outlet) or a specialist but nonscientific subset thereof (industry, technical, business) needs to be constantly borne in mind in considering who the recipients of the information are and what they need to know. This should then inform the selection of language and tools that will be most successful in getting the desired message across to them.

6. Avoid jargon or intimidating language, as discussed throughout this chapter, but yet aim not to be patronizing or speaking down to the journalist or audience. Deploy the strategies mentioned earlier like humor, analogy, and personal anecdote, all of which will help them write a more engaging piece.

7. Don't waffle. Don't bullshit. Keep answers short and concise. It's okay to admit where the researcher doesn't know the answer to a question they have been asked. Some questions asked may appear annoying or even stupid, but that is because the researcher knows the topic and the interviewer does not, and, if the journalist doesn't understand by the end of the conversation (again), that is the interviewee's fault and not theirs.

8. Be available, as press releases lead to an expectation that the researchers concerned are available readily over the next 24–48 hours.

9. Check about expectations of exclusivity, as some larger broadcasters may be wary about scheduling an interview with someone who they might regard as being oversaturated in the media at a particular time.

10. Don't cut off the interviewer by accident during a live broadcast (not that this ever happened to me, and certainly not twice).

Finally, in thinking about the media, I often think of what the great physicist Richard Feynman said when asked by a journalist, the morning after winning a Nobel Prize, if he could summarize in one sentence what he won it for; he replied "Buddy, if I could tell you in one sentence, it wouldn't be worth a Nobel Prize." Not every complex scientific idea or fact can be boiled down beyond a base level of simplicity, but researchers need to work with journalists and the public to guide them gently as close to that cut-off point as possible.

Communication with Other Stakeholders

For many researchers, there may be other stakeholders with whom they need to communicate about their research and findings, besides the peer audience of the scientific community and the general public (either directly or through the medium of the media), or subsets thereof with a specific interest in the research in question (such as patients or farmers interested in research on specific disease or agricultural research, respectively). Such stakeholders might include:

• Industry or others interested in commercial or practical implications of research;

- Funding agencies who have funded, or may wish to fund, research;
- Government or regulatory agencies who have a policy or practice responsibility that may be impacted by a certain area of research.

In all cases, as always, the starting point in preparing for such interactions is to understand who the audience is: where they are coming from; what are their motivations, interests, and goals; and what would a successful outcome of the communication look like. On the basis of these considerations, a message can then be designed and delivered using language and rhetorical tools that are most likely to result in their getting from where they start off to where the researcher wants them to get to, as effectively as possible. In each case, a good basis for preparation could be to list the questions they are likely to ask (why is this important? why is or was this good value for money? what do we need to change because of what you have found?) and make sure the answers are either included in the report or presentation being prepared, or are ready to be delivered with confidence in response to their being presented as questions.

Funding agencies are, for most researchers, a critical constituency as, without their support, research might literally not be possible at all. Researchers will have two main points of interactions with them, when asking for money to do research (the proposal stage) and when accounting to them for what that money, if granted, was spent on. In the book *The Chicago Guide to Communicating Science* (2003) Scott Montgomery referred to every proposal being a *request* (for funding and the agency's interest), an *argument* for the importance of the ideas being presented for funding, a *blueprint* for how the proposed work will be completed, and a *promise* that the researchers can and will deliver the proposed work responsibly and within the time and resources indicated.

Researchers will probably spend almost as much time writing proposals as they do writing papers, as funding is the essential lifeblood without which students and researchers cannot be paid, materials cannot be bought, and great ideas cannot be brought to fruitful conclusion. Success rates for proposals, however, are probably even worse than for getting papers published, and many funding agencies fund less than 10% of the projects for which they receive proposals. The ability to craft good proposals, therefore, is a key skill for researchers and, while space precludes further exploration of that craft here, most institutions will have professional officers to advise and support on this. This is also a key area where experienced researchers can mentor and support new colleagues, by sharing examples, templates, advice, and experience.[21]

At the other end of the researcher-funder relationship, researchers need to be able to write effective reports that demonstrate the key outcomes of the work. In such reports, it is critical to understand the expectations and demands of the audience, which are to know that funding provided was put to constructive and

---

[21] A certain scar-comparing scene from *Jaws* springs to mind here.

responsible use, that timelines and objectives agreed were adhered to (or else good reasons why not are presented) and that the work has had impact, whether through publication, communication at conferences, or (and increasingly) direct communication with key stakeholder audiences.

A key example of such audiences might be industry or commercial audiences, who are interested in what researchers have achieved but, critically, are specifically interested in what potential significance that work may have for their own commercial interests. In my own applied field of food science, a significant proportion of my work relates indirectly or directly to food companies, Irish or international, and much funding has come from such sources, while hopefully much of its impact has been of benefit in that sphere. Today, it is, for better or worse, a general expectation that research has economic impact, and this requires interaction with those who can deliver economic activity, whether through commercialization of research, transfer of technology or know-how to practice, or employment of others. Researchers increasingly need to understand the economic context of their work, which might lead in some cases to their being involved directly in commercialization of their work (through spin-out or campus companies, for example) or indirectly through other companies or organizations realizing the commercial value of the work.

In any case, interacting with industry or other commercial entities relies once again on starting with the audience and understanding as well as possible their motivations and objectives, and then (critically) articulating what the benefits and dangers of the interactions for the researchers might be. Key points here might revolve around the rights to ownership and publication of work and, in my view, a guiding principle for a researcher whose salary and research is largely funded by public (taxpayer) sources is to find a balance between generation of research disseminated as widely as possible (through publication) for open public benefit and achieving the satisfaction of knowing that some of the research has had identifiable uptake for economic or other benefit. I have had several projects funded entirely by industry partners where there was no barrier to eventual publication of the research, but some of the complexity these arrangements can entail will be explored in the next chapter.

In addition, there are cases where research funded by companies does not lead to publication of outcomes, and inevitable questions then arise as to whether the research was not published because the results were not favorable toward the sponsor. For example, if a study on a company's drug finds that it does not work, this may not be an outcome seen as being in the sponsor's interest to publish, yet researchers may be caught in a significant ethical bind if the finding is something they believe the public needs to know.

There is also the complicating factor that journals can be less inclined to publish negative results (that something doesn't have an effect, or that a hypothesis was not verified), and such considerations led to calls (such as from the World Health Organization) for clinical trials to be registered in advance, and findings published (irrespective of their nature) within 12 months of completion; this is becoming increasingly standard practice internationally, and trials that have not been registered are unlikely to be accepted for later publication.[22]

If universities were to become research arms of industry, where every project was locked down by confidentiality and not publishable, this would not be for the good of science or the public, but it should be possible to achieve a balance between different audiences and forms of research to balance the multiple demands for, on the one hand, scientific excellence, and on the other measurable impact.

The last group of stakeholders mentioned earlier is government and policymakers, and again researchers in some (not all) fields may have a requirement or expectation to present their work to such bodies where their findings may have implications for national or international practices. Again, the key is to know the audience and what they want to achieve, matching that against the message the researcher has to share, and designing the most effective strategy to achieve an outcome that satisfies all parties.

There is an unhappy history in some cases, however, of science mixing with politics, and stories of researchers' work being taken out of context to justify decisions and actions that the researchers would not have supported, and so this is a high-stakes activity that needs to be approached carefully and with a high degree of preparation. There are also cases where researchers are put in extremely difficult positions where their work is concluded to be politically inconvenient, and steps are taken that amount to censorship of research that is not consistent with the prevailing political ideology of the time.[23]

To take one example, there is an extensive literature on political interference over many years in the United States with the results of research in climate change. When James Hansen, a leading NASA researcher who developed his expertise studying the atmospheres of other planets, made some very troubling findings when applying such models to Earth, he was assigned a "minder" (whose scientific credentials were at first glance weak, and at deeper glance falsified) whose job it was to monitor and alter every paper, presentation, and press release Hansen generated.[24] It was reported, based on surveys conducted in the period 2005–2007, that, of over 1,700 climate scientists, over 60% reported political pressure, such as demands to eliminate "climate change" from reports or amend related research publications to change their meaning.[25] This highlights the need for positive and open dialogue, and most importantly trust, between government and the scientific community, and the right of scientists to report their findings and provide the best evidence they can, based on their expertise, without fear or favor.

---

[22] It has been required by the Food and Drug Administration in the United States since 2007.

[23] A lot of examples can be found in Mooney, Chris. *The Republican War on Science.* (New York: Basic Books, 2006).

[24] For a good account of this, see Bowen, Mark. *Censoring Science: Dr James Hansen and the Truth of Global Warming.* (New York: Plume, 2008).

[25] https://www.ucsusa.org/our-work/center-science-and-democracy/promoting-scientific-integrity/federal-climate-scientists.html#.W3F9O5NKhR4

The Place of Social Media in Modern Research

In the modern world, it has never been easier to communicate instantly with a massive audience, global in fact, through Twitter, Facebook, LinkedIn, and a host of other social media platforms. As stated earlier in this book, while in theory a researcher could use these communications channels to broadcast their work globally without filters, this would not be a credible communications strategy, as the scientific community would not take seriously anything that has not been quality-assured through the process of peer review.

Of course, there is no reason why such formal peer-reviewed communications cannot be accompanied and complemented by dissemination through social media. Many of the considerations are the same as discussed earlier for more traditional media, but here there are even fewer filters between messenger and audience, and speed is incomparable.

In my experience, researchers who trained and learned their craft before the advent of social media (I am not talking ancient here, as the cut-off could be less than a decade ago), myself included, have not fully grasped the advantages of social media. Such researchers are much less likely to tweet about their research than younger researchers who do not remember the pre-Internet dark ages (but among whom the likelihood of their tweeting about their research is probably still less than about most other aspects of their lives).

Researchers can also blog about their research, or feature professional details on LinkedIn, perhaps keeping this more for professional profiles than Facebook, which covers their nonlaboratory lives. They might also use specifically research-focused social media platforms such as ResearchGate. The different social media systems have different focuses and objectives also, from dissemination (blogs and Twitter) to network building and broad information provision (LinkedIn), and also sourcing information and news from relevant individuals and organizations (also LinkedIn).

There is surprisingly little hard data on the use of social media by scientists. In 2013–2014, a survey of 587 research scientists from 31 countries stated that, at that time, more than 50% used social media (mainly Twitter), and most of these (78%) were 21–39 years old.[26] That study reported that barriers to more widespread use of social media among scientists included a lack of time, a lack of knowledge or confidence with the platforms, lack of clear benefit, and a feeling that such activities were not "scientifically rigorous" for professional communication. A 2017 article in *Nature* reported that the percentage of scientists and engineers surveyed who used LinkedIn, Twitter, and Facebook for research-related purposes was 41%, 13%, and 28% respectively.[27]

---

[26] Collins, K., Shiffman, D., and Rock, J. (2016) How are scientists using social media in the workplace? *PLOS One*, e0162680. (October 12, 2016).

[27] Social media as a scientist: a very quick guide. *Nature Jobs Online*, August 23, 2017.

In chapter 8, I related a case study of communication of my group's work on 3D-printed cheese at conferences, which was the subject of quite a bit of activity on social and more conventional media. Not long after the paper appeared, I got the first email from a journalist for a website looking for a quote about the work, which I provided, and then another couple, but somehow the story or versions thereof (some containing quotes I never gave) started to appear in a wide range of food and technology websites globally (including the Discovery Channel and CNN Business).[28] In addition, I was rather surprised to receive, on Saint Patrick's Day, a text from a friend of mine to the effect that my cheese was trending on Twitter, not words I have often heard. Indeed, the study was briefly catapulted up some list of "hot topics" on Twitter, with comments ranging from the interested to the bizarre ("I like to eat 3D cheese while watching 3D movies") to the dismissive (rather than curing disease X, Y, or Z, guess what stupid Irish scientists are doing).

I did not intervene directly here (only starting to use Twitter myself much more recently), watching instead with a mixture of excitement, bemusement, and outright horror the speculation and commentary that ensued. I can now go back and analyze the events, using a very handy tool called Altmetric, which describes itself as "Bookmarklet for Researchers." From the Altmetrics website, a tool can be downloaded that sits in a browser bar and then, when a paper is found in a database, is used to analyze that article's history of online shares and mentions. When I apply Altmetrics analytics to our study on processed cheese in the *Journal of Food Engineering*, it shows that the paper has an Attention Score (the site's measure of the quality and quantity of online attention a paper receives) of 93, which is in the top 5% of all research outputs, and is the highest score awarded out of 906 papers analyzed from the same journal. It notes that the paper was mentioned by 8 news outlets, 3 blogs, and 19 tweeters (fewer than I found from other searches, probably because most did not directly cite the paper), and that the biggest communities citing were, geographically, the United States and, demographically, members of the public.

Such tools are very helpful in tracking what, in the future, will increasingly represent a series of ripples of influence and impact associated with any paper, beyond conventional metrics like citations.

My first experience of social media that convinced me of its power came when I casually posted on LinkedIn a link to an article in the Irish newspaper *The Irish Times* for which I had been interviewed and supplied a number of quotes. This was on the Saturday of a long weekend (Easter 2018) and I was fascinated to see how, over the next 48 hours, the post was viewed, liked, and commented on by researchers and staff of major food companies all around the world. For me,

---

[28] A useful lesson was that, with the paper we submitted, in the form of Supplementary Material, a short video that one of the students who worked on the project had prepared. While the paper content was covered by the publisher's copyright, this material was not, and so could be shared or short clips or stills extracted for use elsewhere.

this was an eye-opening example of the unharnessed ability of such platforms to reach huge numbers of people, with a range of possible professional outcomes that might arise from their awareness of a researcher's interest, activity, or expertise in a particular area.[29]

There have been a number of cases where Twitter has provided a medium for very fast moving and sometimes highly powerful debate about science or papers, such as a debate about a paper from NASA in 2010 that reported the discovery of microscopic life forms that could apparently base their metabolism on arsenic, a material famously toxic to most organisms. The key significance of this finding related to its possible implications for development of life on other planets. Huge criticism and demolishing of the findings followed on Twitter (#arseniclife[30]) and various blogs, to which the NASA researchers responded initially by saying they would not engage in what they characterized as an improper way to engage in a scientific discourse.

Initial responses on Twitter were in many cases positive (#thischangeseverything), but turned to increasing skepticism and questioning of the methods used and conclusions reached (several of the most prolific tweeters were journalists). The subsequent loss of interest and belief in the report can be tracked through time through Twitter, following in real time the focus of the scientific community in the topic. A blog post 6 months after the NASA publications reignited the interest, however, as a Canadian researcher who had been particularly critical of the study accepted that there was a need to test and refute the claims made, and eight critiques of the work were subsequently published.

As this example of the role of social media in rapid-fire postpublication peer review shows, it almost goes without saying that social media has a place in the research communication landscape, but researchers need to recognize its inherent limitations (particularly the lack of filters and quality control) and equally its power, in terms of scope, speed, and reach. Combined with the potential to use it for building networks, disseminating information, and learning about events, jobs, and other opportunities, it appears clear that even the most established scientists need to become familiar with the rudiments of its use, lest they end up stuck with the pointy end of a harsh hashtag.

---

[29] I have since applied LinkedIn reasonably regularly for research-related updates on publications of developments in my research group, and consistently found its reach to be unexpectedly wide, and a great way to keep in contact with former students and contacts. I also increasingly use Twitter today (@akellyucc) for shorter updates and observations, and find it to be a platform where one can express a broader profile than LinkedIn, combining the professional and the more personal, as I believe in the idea (#AlongsideScience) that it is important to show that scientists have interests outside science (and so I tweet about topics from recent papers and science news stories to humor, pop culture, politics, and occasionally my dogs).

[30] One study of the affair recorded 35,860 posts on the matter in a 16-month period, 48.2% of which were skeptical of the findings. Yeo, S. K., et al. (2017) The case of #arseniclife: blogs and Twitter in informal peer review. *Public Understanding of Science*, 26, 937–952.

10

Communication across a Career in Science

The changing relationship between researchers and the scientific community, and their standing within their field, means that the expectations, opportunities, and means of scientific dissemination change over a career focused at least in part on research. Consideration of these changes gives a useful insight into the evolution of a typical (if there can be said to be such a thing[1]) career in science, and will be the focus of this chapter.

Getting Started: A Simple Decision to Make with Long-Term Impact

One of the first decisions a researcher needs to make when starting to publish, or preparing to publish their first paper, is a deceptively simple one: what is their name?

This seems like a rather easy question, and one that most of us have been able to answer since we were around a year old. However, when considering a publishing name, a key question is the uniqueness of the name. To look at it another way, what is the chance of someone, at some point in the future, looking for an individual's research or papers, through a database search for example, and not being able to work out which of several identically named researchers is the one they seek.

I remember noting when I was drafting my first paper that, in all the other papers I was reading, people tended to use initials after their first name, perhaps because that looked more official or important, so I did the same. I then continued to do so, and it was some time before I realized how significant and lucky that move was.

---

[1] In this case, "typical" means a researcher working in a university or research institution, but of course researchers can change institutions, roles, sectors, countries, continents, and fields, or move between roles in the public sector, such as a university, and the private sector, such as industry.

If I do a search today on Web of Science for "Kelly, A" (no other search terms besides my name), I find over 4,800 results, but if I search for "Kelly, AL" (including L for Louis), I find just over 300, much closer to my actual total, but including a small number of papers by other AL Kellys.[2] If I had gone to my third initial, O (don't ask!), I would have no overlap with anyone in the database, but AL seems fine to me.

For someone who is publishing their first paper and knows that their future career is highly unlikely to be in scientific research and publishing, this is not a major decision. On the other hand, for anyone who, when starting out, thinks there is even a small chance that this is the start of a long list of papers to come, then this is not a decision to be taken lightly. A quick database search will give a good idea of likelihood of having a unique publishing name when using 0, 1, 2, or more initials, and should guide a researcher in the right direction. Obviously, there is always a chance of someone with the same name starting to publish AFTER you, but aiming for likely uniqueness, even if no one else has the same name now, is still wise. I have had two PhD students with no middle initial for some reason, but their surnames were reasonably unusual; if this were not the case, selecting a suitable initial and sticking to it would be perhaps unorthodox but surely a pragmatic solution.

The name under which a researcher publishes is (to use an imperfect but useful analogy) like a brand. This name becomes intrinsically associated with their reputation, for better or for worse, as credit accrues or blame, in cases of future bad behavior, ensues. Researchers, who frequently suffer from at least mild egotism, like to think they are "someone" within their field, and that their name has a meaning and standing. So, identifying a form of their name that is going to be distinctive and unlikely to be confused with others makes sense.[3] It wouldn't be good to be constantly confused with someone of the same name who had several papers retracted for fraud, for example! In addition, searches by others for names to evaluate their track record or h-index for the many forms of external scrutiny to which a researcher is regularly submitted (for grants, jobs, awards) are greatly facilitated by selection of a unique name, although new systems such as the ORCID coding, which attributes a unique identifier to all authors, will probably alleviate such concerns in the future.[4]

---

[2] Including my favorite scientific namesake, who published on sea lions and snow leopards in the 1950s and 1960s, including a paper on sea lion milk in the year I was born, which seems like an odd yet auspicious omen given my later core research direction (or else a sign I got started really, really early, but I am still studying the first food material I ever encountered).

[3] When publishing my first book, I decided not to use an initial, and of course now search engines are taking their karmic revenge by confusing me for some time with an Irish politician of the same name, who they have decided wrote my book. That soft thud was the sound of my case resting.

[4] The ORCID (Open Researcher and Contributor ID) system gives researchers a unique alphanumerical identifier on an international database, but most people are still going to think in terms of names rather than numbers.

Once a researcher selects their name for publishing, they are then unlikely to change it in the future,[5] as switching to a different name means that, essentially, two different individuals will appear on the databases, and uninformed others are not likely to make the connection between these.

Early-Career Communication Priorities and Needs

When starting to publish and disseminate their work, a researcher will have far more questions than the experience with which to answer these, which is one of the reasons that the currently widespread model of apprenticeship training for early-career researchers is so vital.

In most cases, a researcher at the start of their career will be undertaking a graduate (or postgraduate, depending on country) program such as a PhD, doctoral, or masters degree. They will be completing the major research component of this under the supervision of an experienced researcher, who provides guidance on the planning, conduct, analysis, and (critically) dissemination of their work. The idea is that, by associating with an experienced researcher (or panel of such supervisors, depending on the precise model followed), the neophyte will absorb key practices and learnings so that, by the end of their degree, they have developed a level of experience and independence, and can essentially take off their metaphorical "learner's plates" and undertake research on their own.

Some of the key areas where supervisors can advise in terms of publication are choice of journal, achieving the required standard of scientific writing, navigating peer review and many more areas discussed in this book. In reflection of the input of supervisors, their contribution is generally recognized (where genuinely earned in terms of meeting the criteria discussed in chapter 3) through coauthorship of their students' papers, which helps their own career, as discussed later. It could also be argued that students benefit from the names of these senior researchers (and the reflected glory, however modest, of those individuals' reputations) appearing on their papers.[6]

The broad international understanding of the nature of the PhD degree is that it is awarded to students who convince a panel of expert examiners (typically of international standing) that they have made an original contribution to knowledge in the field of their thesis. The manner in which an original contribution can be deemed to have been made is typically by viewing the work described in a thesis through the prism of publication, as obviously this is the way in which any researcher demonstrates their contributions. Their field in turn validates the work through peer review.

---

[5] Perhaps for reason of marriage, or entering the Witness Protection Program.

[6] James Watson apparently preferred not to have his name on his students' papers on the basis that it overshadowed their own contribution, although apparently some such students believed the benefit of that association outweighed any such downside.

For this reason, a question often asked of examiners of a doctoral thesis is some version of whether the work is "publishable in whole or in part, as a work of serious scholarship." In some institutions and countries, the thesis is essentially submitted when a suitable number of papers have been published in respected journals within the relevant discipline. This number may often be three or four and, in some cases, the equivalence between publication of four papers and the submission of a thesis is quite explicit. In my university, we would encourage students to aim for 4–5 papers, at least in my own area, and, while these could be included in the thesis as draft manuscripts (although having at least some published will greatly reassure the examiners as to the standard of the work), the key question remains whether the work is overall of a standard that it is likely to be, in the future, published in good journals.

So, for a young researcher, publication of good papers in good journals is unquestionably a key objective. Quantity will never override quality, of course, and a small number of papers in very good journals will be more valued than many in lesser ones.

On graduation, a researcher will be ready, if staying on an academic or research-focused track, to apply for their first (relatively more) independent position, as a postdoctoral researcher or research assistant. In applying for such a post, one of the first things the selection panel will view is their list of publications to date, another reason why publishing during the research degree is a key goal, even if a university doesn't require this.

I also encourage research students to think about writing and publishing from the very start of their studies, for several reasons:

Writing an early review of relevant literature is a great way for a researcher to familiarize themselves with an area (much more than reading papers alone, as having to write on a topic forces a much deeper level of engagement with the material), as well as getting feedback on their writing from their supervisor(s) and starting to improve immediately, where necessary. It also allows the supervisor(s) to see how a student understands the area, and whether there are gaps or issues. It is hard to underestimate the psychological benefit of having a folder labeled "my thesis" with actual pages of text, no matter how preliminary, in it from a very early stage, and it is then an easy task to go back to this on quiet days and update and extend this, or perhaps consider submitting it at some stage as a publication in a review journal;

Writing a draft of a research paper based on emerging results can be a great way to see how a piece of research is coalescing, and whether it is complete or coherent, while also catalyzing feedback and analysis from other members of the research team. It is also an ideal thing to do on quiet days in research when nothing else can be done for a myriad of reasons (e.g., the machine is broken and we are waiting for the repair, the ordered chemical is late) and, while there are always days in research at the end of which nothing concrete or useful has been generated, a day of writing will always lead to progress toward a thesis or paper, even in terms of a count of words on a page;

Publishing in advance of submitting a thesis, as mentioned earlier, is a great way to convince the examiners that the work is of high standard. If experts in the field have passed it in the process of peer review, that should be a significant reassurance of quality, although of course in every case the thesis will be examined on its merits nonetheless. All material is open for criticism and questioning by the examiner and defense by the candidate to demonstrate that they have complete ownership and understanding of the subject (and did not just follow the instructions of their supervisor(s) without their own intellectual input);

Publishing before completing a degree is a critical step in ensuring that the research contained in a thesis is disseminated at all. I am convinced that the libraries of universities are full of great unpublished work because the student in question got a job (and became too busy to get around to publishing) while the supervisor(s) also did not have the chance or sufficiently close relationship with the data to do so. There are also cases where a student may leave thinking that publications are not a priority in terms of their career goals, but perhaps later regret not having those on their CV if they wish to apply for a post for which they would be an asset, with the time since completion then being too long to realistically publish work that would be seen as current.

In addition to publication, early-career (during and immediately after their degree) researchers need to broaden their dissemination strategy in terms of "building their brand" and extending their reputation and network of valuable contacts within the field. This is principally achieved through participation in conferences and seminars, as discussed in chapter 8.

A typical place to start such contributions will be a poster, which as stated earlier is pretty much the lowest level of prestige in publication, but gives someone the excuse to be at a conference. It can be a catalyst for putting down the first flag of "this is me, and this is my research" and perhaps making some initial contacts, while getting feedback through the comments and questions of those who will view it.

Next up the rung on the achievement ladder will be the presentation, whether at a local, national, or international conference, which is a key opportunity to present work, build reputation, raise personal awareness, and hopefully establish helpful links with others. Very early-career researchers should also seek any possible opportunity to contribute in different ways to conferences, for example by volunteering to help out with administration, standing on information or registration desks or running around with microphones during question-and-answer sessions; such jobs can be a great way to get a free or discounted registration for a conference, and get in the door and up close to speakers and other contributors.

All in all, in the early stages of a career, a researcher needs to do all the hard work, in terms of writing the papers, handling the responses to peer reviews, submitting abstracts and requests to go to conferences, and then, if successful, preparing the talks and posters. This is all hard work, but should be reflected in the recognition that comes from the prized first authorship on all of the above, as this is likely to be the key part of a career where this is associated with almost every communication (early collaborations with other students or researchers on projects could lead to some second-place authorship).

Research papers will be the key output of this phase, and perhaps a review article or book chapter could be leveraged from the literature review undertaken for the thesis.

Of course, some projects could have nonpublishable outputs of high value to specific stakeholders, such as patentable material or that which a sponsoring company has indicated it prefers not be published for reasons of disclosing intellectual property or trade secrets, at least not immediately. In such cases, any restrictions of potential to publish should be agreed up-front by all parties involved, and the supervisor(s) and student must deem that the benefits of this arrangement (in terms of career options, impact, or other factors) outweighs the constraints proposed.

I once had a project sponsored by a company where there was a clause in the agreement that publication was allowed and encouraged once no commercially sensitive material was disclosed; the study was undertaken on a model system that related to but was independent of the company's key food product of interest, such that the results would be useful but generic enough to be publishable. However, when we went to publish papers based on the work, I sacrificed many hours of my life to meetings with lawyers arguing about whether details about their product could in any way be magically guessed based on details included in the Materials and Methods sections of the draft papers; several papers were eventually published, but this was a clear and time-consuming example of the complexity that can arise if any such restrictions and their application in practice are not explicitly agreed and understood by all involved.

As explored in chapter 9, this is also a key period in which researchers should seek additional opportunities to develop communication skills and a researcher's contacts and reputation through nonspecialist dissemination, such as newspaper articles, local competitions, or research outreach programs. Today, many early-career researchers will far outstrip their supervisors in understanding the power and reach of whatever the key social media platforms of the day are. They might establish blogs or even tweet about their research, or publish articles in nonspecialist online journals or websites, which as mentioned earlier can sometimes lead to the most widely read outputs of their research. At a minimum, researchers at this point should establish a LinkedIn account to start building their online profile, and start to use media such as Google Scholar.

Getting Established: Priorities and Needs as Early-
Transitions to Midcareer

Following the completion of a research degree, and perhaps some initial stages as a
postdoctoral researcher, a researcher who is building a career in research (unless in
an industrial or other private-sector environment) will move into a new phase where
their relationship with different kinds of dissemination of research will change.

In such a phase, they will cease to be a supervisee, and become a supervisor, ideally securing research funding such that they can recruit and mentor students of
their own, inevitably influenced, either positively or negatively,[7] by their own experiences to date. One key change here is that their place on an author list might jump
from near the front to near the end (which has its own importance, as we have seen).

One significant red-letter day that might happen around this stage is the first
time an invitation to peer review a paper for a journal is received by a researcher, as
it means that someone (and not just anyone, but a journal editor!) (1) has heard of
them and (2) is interested in what they think. This is a good sign that a researcher
is making progress, and they should grab any such opportunities gladly to build
experience and relationships with journals, but perhaps seek advice from others on
how best to approach the task.

A researcher at this stage will probably not be presenting posters at conferences,
but encouraging their students to do so, and helping them with the preparation.
They might be presenting at conferences on their own (or their students') work,
or encouraging students to do so, and perhaps might receive some invitations to
present at conferences, another key metric of progress.

They may also be increasingly writing review articles (sometimes ones they pro-
pose, sometimes ones they are invited to write) for journals or books, on the basis
that that their active role in that area of research gives them a useful perspective on
key activities, findings, and knowledge gaps.

For those researchers in an academic environment, these activities will likely be
combined with establishing a teaching program, developing courses and modules,
and disseminating knowledge in a different context. This will hopefully include
advanced courses, which build on their ability to talk about their fields of research
from the perspective of someone directly involved at the coalface of knowledge
creation in the topic in which they are teaching, in the spirit of "research-led (or -
based or -informed) teaching."

Giving Back: The Roles and Responsibilities
of Senior Researchers

As a researcher matures, and moves into the phase that begins perhaps 20 years
post-degree, they are likely to be in a senior or professorial role in their institutions.

---

[7] As a great philosopher once said, "If you can't be a good example, you'll have to be a terrible warning" (actually, I saw it on a fridge magnet).

Their publication lists will have grown, and will be one key reason they have gotten to the position they now occupy. On their papers, they are likely among the last-named authors (as supervisor) or perhaps on other papers somewhere in the middle, as they become part of larger teams thanks to national or international collaborations built up with other researchers or institutions.

They will be regularly invited to review papers for other journals, but they hopefully have not forgotten their responsibility to the community. As one wise US colleague once put it wisely to me (thanks Allen!), we spend the first half of our careers benefiting from the work and advice of others, and we should spend the second half paying that back.[8] As a simple example of this, years of sending invitations to others to connect on LinkedIn will give way to regular requests from others to connect to them.

Senior researchers will also have the opportunity to become editors of journals, and such invitations will be extended based on their reputation and contribution to the field, and will be accepted based on consideration of the benefits and importance of the role (they definitely won't do it for the money!) balanced against the commitment of time and energy required to do it properly. Becoming an editor gives a researcher a key opportunity to become an architect of a field, deciding what gets published, how, and by whom. To me, this has been a hugely rewarding part of my work over the last almost 15 years, but to do it properly is not a casual undertaking.

In terms of conferences, there are likely many and frequent presentations and posters given by the researcher's students, but personal presentations will likely be keynote overviews to introduce sessions in most cases. Such invitations will usually come with financial support to attend the conference, leading to many opportunities to travel and broaden experiences and networks of contacts. Other conference contributions might include chairing sessions, being invited to be on scientific committees for conferences (with responsibility for selecting which submitted abstracts will be accepted for presentations), or even going as far as to organize conferences directly.

I clearly remember once reading a book while on vacation about the development of a particular field of medical science, and I remember vividly where I was when I read about someone deciding that a key step to advance the field while it was growing was to organize a conference and bring everyone in the area together while it was still possible to do so. I decided on the spot to do the same for an area of my research (milk enzymes) and around two years later hosted the first international conference on the topic, with around 150 of the key active researchers on the topic from around the world, in my university; this was hopefully a useful catalyst for new conversations and collaborations on the area which attendees still refer to when I meet them.

Senior researchers may also have the opportunity or invitation to participate in books as editors, either through invitations to do so, or through deciding (alone

[8] Although I do not think that every senior researcher behaves this way, unfortunately.

194   How Scientists Communicate

or with colleagues) that there was a gap in the market for a book on a particular topic, and then approaching a publisher with a proposal to do so. A key consideration here is that the proposed editor will have a sufficient standing and network of contacts in the topic of the book to be able to assemble a coalition of willing authors who will produce incisive articles on their subspecialties.

A further key step may be a decision to complete a sole- or coauthored book, on an academic or nonacademic topic, a commitment of significant time and energy compared to any other form of communication activity.

All these communication activities may also be needed to be balanced against other commitments of an academic or research role of increasing seniority, such as management roles (dean or head of department, center, research group), teaching responsibilities, sitting on committees, and grant submission and review. Individual researchers must constantly balance the relative attention and energy they can devote to the different demands on their time.

In my own case, I would say I spent a decade as an early-career researcher, then combined midcareer with a decade of successive nonresearch leadership roles in my university, and then emerged with a specific decision to focus again on research (in a more senior role) while devoting time to a long-held desire to produce forms of communication that I had never had time to work on to date; you are holding one product of this desire in your hands now.

One other factor that will gradually change over the course of a career in science is a researcher's standing and esteem within the field, and how they and their work are objectively viewed and evaluated by others. Some consideration of this will be driven by the ego that will always be found in researchers who have even modest pride in their work and achievements, but it is also critical in terms of making claims for standing when trying to impress bodies such as grant-awarding agencies or appointment or promotional committees. One of the key modern metrics in this regard is the h-index.

The H-Index

We know by this point that citations are the key currency, rightly or wrongly, by which scientific quality and impact are most commonly measured. This applies both in terms of papers (more citations indicates that a paper has had more impact in terms of influence on subsequent work) and of journals (for which the impact factor or other scores aggregate data around the relative proportions of citations to papers published compared to numbers of papers published).

What about the Equivalent for an Individual Researcher?

To consider why evaluation of citation performance is important, consider the perspective of a senior researcher reviewing a pile of CVs from applicants for a research position in their institution, or a funding agency considering a number of applicants for a grant to support a research project. How do they short-list from a large number of applicants to a smaller set, and ultimately to the successful one?

Obviously, for any job or grant award there will be all sorts of criteria, and fit of experience to the opening in question in terms of these criteria will be a primary determinant, but every such evaluation will also involve some form of evaluation of the overall scientific credibility of the individual in question.

There are many ways in which this could perhaps be judged. In predatabase days, when citation data-crunching was a far less automated and accessible process, a simple and crude judgment might be made on the basis of the length of the person's publication list: lot of publications good, few publications not so impressive. However, this does not take into account the quality of the publications, the standing of the journals in which they appeared, or their significance within the field.

Always, Quality in Science is Far More Important Than Quantity.

Any experienced evaluator, even before Impact Factors became readily available, would know the relative standing of journals in a particular field, and would know whether publications on the list had appeared in respected international journals, or journals of local circulation, poor standing, or questionable editorial or reviewing standards, so this is the next level of scrutiny that might be applied. As discussed earlier in this book, it is a fair assumption that a journal of higher standing applies higher reviewing standards and has higher rates of rejection. So, to have got a paper into a high-impact journal earns it (and by extension its authors) a greater degree of respect and credit than one in a less prestigious journal (noting again that what constitutes a "high" Impact Factor can differ widely between fields, as the range of Impact Factors for biomedical science journals stretches a lot higher than that for food science, for example).

Another key criteria by which quality is judged is the manner in which a researcher's papers have attracted citations. A list of papers in a CV might then indicate for each how many times they have been cited. This is very useful information, and a broad portfolio of well-cited papers in good journals (relative to the norms of a field) will make an impressive CV.

So, if we wanted to summarize an individual's citation-related prestige in one number, we could perhaps just add up the citations to give a total citation count, but this might reward someone with a few "megahit" papers (like a very helpful and widely used method, as illustrated in chapter 6) or review articles (which tend to attract more citations than research papers). Any calculation of average citation levels (total number of citations divided by number of papers published) would also be flawed in that it rewarded lower productivity (giving a smaller number by which to divide the citation total). Instead, the leading metric that has emerged to summarize in one number publication level plus citation history is called the h-index, which was first proposed by a physicist called Jorge E. Hirsch in 2005.[9]

---

[9] Hirsch, J. E. (2005) An index to quantify an individual's scientific research output. *Proceedings of the National Academy of Science of the United States of America*, 102, 16569–16572.

A h-index of 10 means that a researcher has published 10 papers that were cited at least ten times. They might have published 50 papers, but to evaluate the h-index these are ranked in decreasing number of citations, and then counted from the most-cited backward. The point at which the number of papers equals the number of citations is then the h-index. In this example, 10 papers were cited at least 10 times, while the other 40 papers were cited less than 10 times.

In the original paper, Hirsch proposed a second index called the m-index, which was calculated by dividing the h-index by the years since the appearance of the first publication counted. This always seemed very fair to me, as an h-index will inevitably increase with a researcher's age, and achieving an h-index of, say, 20 in 10 years is to me a more impressive achievement and reflection of impact on a field of research than reaching the same value in 30 years. However, this time-corrected parameter has never caught on.

Many citation and publication databases now calculate h-indices for researchers based on the citation data they track, but interestingly (and significantly) they do not all agree on the same h-index. For example, I have quite different h-indices on Google Scholar, Scopus, and Web of Science (presumably because of different bodies of publications tracked, and inclusion or exclusion of things like book chapters), and, while someone will obviously be tempted to use the most favorable accurate h-index, it is critical to state where this was obtained from.

This is vital because the decisions increasingly being made on the basis of these numbers are hugely influential for a researcher's career. For example, a CV of an experienced researcher will be expected to declare their h-index explicitly, while some funding agencies will actually set minimum values for h-indices for applicants for their most valuable funding instruments. The h-index can also be used to calculate the citation impact of groups of individuals, institutions, or even countries. Considering it has only been in use for under 15 years, it is astonishing how widely the h-index is used, including for key decisions on hiring and funding.

It has been pointed out that certain prominent scientists would not be flattered by application of the h-index calculation, most notably Einstein's if he were assessed before around 1910, by which time he had only published a small number of papers, but those had completely overturned physics. Other limitations of the h-index include the fact that it ignores an author's place on an author list (i.e., there is no difference between being first, last, or buried in the middle), increases but can never decrease (thus favoring older or more senior researchers rather than younger, or at least more early-career, academics), and ignores citations to papers beyond that level required to achieve a particular h-index. For these reasons, Hirsch himself has noted that the h-index should only be considered, when making major decisions, alongside other forms of evaluation of a researcher's impact and output.

Proposed variants to the h-index include a h-index normalized by the number of authors on the papers counted, versions that take into account an author's place on an author list as a form of ranking, and some very mathematically complex formulae that take into account factors such as collaborative distance, and nonentropic distribution of citations (nope, me neither). A very simple alternative to the h-index proposed by Google and available with Google Scholar is the i10 index, which is simply the number of publications an individual has with more than 10 citations, but this remains little used.

The Curious Case of the Erdös-Bacon Number

Besides the h-index, some fields have their own unique ways of recognizing prestige or standing. In the case of mathematics, one very unusual such parameter is called the Erdös number, which is named after a Hungarian-born mathematician called Paul Erdös (1913–1996), who spent much of his life essentially homeless and stateless. He had a habit of arriving on the doorsteps of other mathematicians and declaring "my brain is open," which meant that he was coming to stay and work. In that period, he would do nothing other than discuss maths with his host, besides sleeping very briefly and consuming vast amounts of coffee (he once said that a mathematician is a machine for turning coffee into theorems) and perhaps amphetamines.

He published over 1,500 papers in his lifetime, with a huge number of coauthors, and today mathematicians relate to him, as if he were the center of their field around which all others orbit, through the Erdös number. An Erdös number of 1 is awarded to someone who published with the man himself, a value of 2 indicates someone who has published with someone who published with Erdös, a value of 3 extends the chain of relationships one more arm's-length away, and so forth. It has been estimated that 90% of the world's active mathematicians have an Erdös number less than 8, and what greater tribute could there be to such a key figure in any field than that others still remember him and pride themselves on even the most tenuous link to his greatness.

Other fields now have similar types of indices, particularly in physics, where numbers of steps to reach back to luminaries such as Erwin Schrodinger, Wolfgang Pauli, and of course Einstein, or more recent figures such as Peter Higgs (of the boson) are sometimes quoted.

Perhaps more familiar in popular culture is the Bacon number, where actors or others were linked into terms of number of steps of linkage to the actor Kevin Bacon, apparently because of the large number of films with large casts in which he had appeared. Intriguingly, however, there is such a thing as the Erdös-Bacon number, which is the sum of steps to both central figures named, suggesting a closer relationship between the worlds of Hollywood and mathematics than perhaps might be guessed. The late Stephen Hawking had an Erdös-Bacon number of 6, having a Bacon number of 2 and an Erdös number of 4, while the actors Kristen Stewart, Natalie Portman, and Colin Firth have an Erdös-Bacon number of 7, arising from them all being listed as coauthors on papers to which they were linked.

In conclusion, any researcher needs to have a wide range of skills and techniques at their disposal for communication, but the ways in which they deploy these, the circumstances and objectives of their communication, and the platform from which they can influence and contribute to their field, all change through their career. The two pillars that run through any scientist's life are (1) doing research and (2) communicating about it, both during the activity and afterwards to share the outcomes. As a career progresses, the relative proportions will likely change, and it could even be argued that, for that career to progress in the most successful way possible, communication skills and communication-related activities are the most important factor.

{ 11 }

Some Final Thoughts

I started this book by boldly claiming a key role in civilization for the scientific paper, as the means by which we record and pass into common knowledge and application the findings that extend our knowledge of life, our planet, and the universe. Science has created around this principle an architecture to ensure that only those papers that bear findings that are apparently credible and honest achieve the status of publication and acceptance by the broader community, by virtue of their having passed, in theory, through the intense scrutiny of some of the best minds in the field.

There is only a small proportion of humanity who are prepared to devote their life to research, finding that this activity fulfills a certain need in their nature. The narrow size of this slice of the population reflects the fact that successful scientists must possess certain characteristics like curiosity; tenacity; a love of puzzles, problems, and challenges; an ability to handle uncertainty and defeat; high optimism, self-confidence, and belief; the ability to work well both individually and in a team; excellent organizational, planning, and communication skills; strong qualities of honesty and personal integrity; and perhaps a bit of madness too.

Scientists are trained in a particular way of doing research, typically at first though apprenticeship and osmotic transfer of craft from experienced researchers, and learn above all else that their work will be complete only when it appears in a good paper in a respected journal. They know that their research is not complete until it has been recorded and passed on to those for whom it will have significance or be of impact, whatever that may be.

They also learn that, to progress in their chosen vocation, they will be judged, almost above all else, on the quality and quantity of such papers. Thus, they curate and nurture their publication lists like prized gardens, counting ephemera like citations with an obsession derived from the knowledge that these will be used as indicators of their success, and thereby validate their life's work.

They are members of a niche community, first of researchers but more specifically in their own specialization, and, on being accepted as part of such a community, they accept certain responsibilities. These include sharing in the self-policing activity of peer review, which seeks to keep the literature pure and untainted by substandard or questionable research.

Scientists must also hold themselves, and others, to the highest standards of personal and professional integrity, as they are offered the privilege of being trusted in what they say and write, as to mistrust everyone would require a level of checking and suspicion that would cause the system of science to both grind to a halt and tear itself apart. Trust comes with responsibility, though, and the most severe of repercussions must befall those who abuse such trust, for wasting the time, money, and respect of others. Of the penalties that result from such infringements, perhaps the most damning is damage to the most important thing to many researchers, which is their reputation and their good name.

Scientists do not just toil in their laboratories and offices, their existence betrayed only by a steady flow of papers bearing their name, but also are engaged in a fundamentally social world, in which they are expected to present (and benefit from presenting) their work at conferences, in forms that are more transient than their papers but serve a different function, more catalytic and interactive, like oil that keeps the gears of thought, collaboration, and learning in the machine of science moving constantly and smoothly.

Today, they also need to master a wide range of communications skills and strategies, and be prepared and able to speak not only to their expert peers but also to audiences from schoolchildren to the general public and the media, and target the level and means of communication accordingly.

For these reasons, it can be argued that perhaps of all the qualities of a scientist just listed, the most important is excellent communication skills. I fully believe that, to be a successful researcher, it is critically important to write, speak, argue, and publish well, and that those who are excellent in the laboratory but cannot engage in open discussion and interaction with their peers will get to a certain level, but not ascend to the key leadership (thought or organizational) roles that represent the pinnacle of a scientific career.

Science doesn't exist without communication. The scientist who makes a key discovery and keeps it to themselves, or communicates it so badly that their audience does not understand or appreciate the meaning and importance of what they find, stops fatally short of their responsibility to add to the storehouse of human knowledge.

In this book, I hope I have given a sense of the complexity, responsibility, and practices of modern scientific communication (at least as it stands in 2019, an important caveat, given the pace at which this area of activity is evolving). Above all else, I hope I have communicated the vital significance of being able to communicate, often and well.

The philosopher René Descartes famously said, "I think, therefore I am."
A scientist could say, "I do science, therefore I communicate."
What does science produce?
It produces knowledge.
Well, what does knowledge look like? Isn't it abstract?
Yes, well it has an abstract, and usually starts with a title, followed by . . . .

{ BIBLIOGRAPHY AND NOTES }

The following books have all been helpful to me in different ways in my writing and learning how to communicate, or just how to be a scientist. Not all are books by or about scientists, so apologies if the list is a little eclectic!

Aldersley-Williams, Hugh (2005) *Findings: Hidden Stories in First-Hand Accounts of Scientific Discovery* (Lulox Books) [Takes several key papers from the last century and explores how they are written, especially the the use of language and argument].

Bryson, Bill (2004) *A Brief History of Nearly Everything* (Broadway Books) [Simply a masterclass in scientific writing for the general public].

Carpenter, Humphrey (1977) *JRR Tolkein: A Biography* (Harper Collins) [Included simply because it, more than anything else (and in particular a chapter called "Oxford Life)," made a very young version of me want to be an academic who could write, and was a profound inspiration for me throughout my life].

Collins, Harry (2017) *Gravity's Kiss: The Detection of Gravitational Waves* (MIT Press) [For its detailed exploration of how a 1,000-author paper comes to be written].

Day, Robert A., and Gastel, Barbara (2011) *How to Write and Publish a Scientific Paper, 7th Edition* (Greenwood) [A classic very comprehensive overview of scientific communication, covering a lot more than writing and publishing a paper, and full of useful tips and examples].

Deck, Jeff, and Herson, Benjamin D. (2011) *The Great Typo Hunt: Two Friends Changing the World, One Correction at a Time* (Broadway Books) [You will laugh, but you will also think more about the danger of badly used punctuation, and every editor has probably felt some kinship with these authors and their quixotic quest].

Friedberg, Errol C. (2005) *The Writing Life of James D. Watson* (Cold Spring Harbor Laboratory Press) [Because there are very few books that focus on the specific writing styles and habits of famous scientists, as this one does].

Gordon, Michael (2017) *Scientific Babel: The Language of Science from the Fall of Latin to the Rise of English* (Profile Books) [A very good history of how English came to dominate as the language of scientific communication].

Holton, Gerald (1998) *The Scientific Imagination* (Harvard University Press) [Includes some very nice examples of the relationship between scientist's processes and their final papers, including a very good discussion of the oil drop experiment of Robert Millikan].

Judson, Horace Freeland (1996) *The Eighth Day of Creation: Makers of the Revolution in Biology* (Cold Spring Harbor Laboratory Press) [For a great example of how to make a huge scientific historical account immensely readable].

Judson, Horace Freeland (2004) *The Great Betrayal: Fraud in Science* (Harcourt) [A thorough overview of, and discussion of, a range of cases of fraud and fabrication and other ethical issues in science].

Kevles, Daniel J. (2000) *The Baltimore Case: A Trial of Politics, Science and Character* (W.W. Norton & Company) [A detailed but readable account of the saga of the investigations into a single 1986 paper from the laboratory of Nobel Prize winner David Baltimore].

King, Stephen (2001) *On Writing: A Memoir of the Craft* (Simon & Schuster) [A great insight into the power and craft of writing, albeit not quite in the scientific field, but still containing many thought-provoking observations].

Kirkman, John (1992) Good Style. *Writing for Science and Technology* (Longman) [An excellent overview of key principles of good scientific writing].

Leith, Sam (2012) *You Talkin' to Me? Rhetoric from Aristotle to Obama* (Profile Books) [An entertaining introduction to basic concepts of rhetoric and argument, a handy part of the communicator's toolkit].

Lightman, Alan (2006) *The Discoveries: Great Breakthroughs in 20th Century Science* (Vintage) [Takes many classic papers and, including some or all of each paper, discusses its historical context and how it is written].

Montgomery, Scott (2003) *The Chicago Guide to Communicating Science* (The University of Chicago Press) [A useful overview of many aspects of scientific communication].

Overbye, Denis (1991) *Lonely Hearts of the Cosmos* (Picador) [A beautiful overview of cosmology and particle physics, recommended here as a great example of science writing for a popular audience].

Schecter, Bruce (2000) *My Brain Is Open: The Mathematical Journeys of Paul Erdös* (Simon and Schuster) [A very entertaining biography of a famous figure, which explains the Erdös number very well].

Strunk, William, and White, E. B. (1999) *The Elements of Style (Fourth Edition)* (Longman) [In terms of the impact per gram of book, this tiny volume packs more impact on writing than any other book I could name].

Truss, Lynn (2006) *Eats, Shoots and Leaves: The Zero Tolerance Approach to Punctuation* (Avery) [A book like this gives a clear appreciation of the importance of good grammar, without actually being a book about good grammar].

Watson, James (1980) *The Double Helix: A Personal Account of the Discovery of the Structure of DNA* (W.W. Norton & Company) [I own several editions of this book, and this 1980 one includes commentary, contemporaneous reviews of the book and reproductions of original papers relevant to the text. I also like the 2012 Annotated and Illustrated Edition (Simon & Schuster), which includes many additional images and appendices].

{ INDEX }

For the benefit of digital users, indexed terms that span two pages (e.g., 52–53) may, on occasion, appear on only one of those pages.